Yoshiyuki Sakai

Coherent structures and secondary motions in open duct flow

Dissertationsreihe am Institut für Hydromechanik
Karlsruher Institut für Technologie (KIT)

Heft 2019/1

Eine Übersicht aller bisher in dieser Schriftenreihe erschienenen Bände finden Sie am Ende des Buches.

Coherent structures and secondary motions in open duct flow

by
Yoshiyuki Sakai

Scientific Publishing

Karlsruher Institut für Technologie
Institut für Hydromechanik

Coherent structures and secondary motions in open duct flow

Zur Erlangung des akademischen Grades eines Doktor-Ingenieurs
von der KIT-Fakultät für Bauingenieur-, Geo- und Umweltwissenschaften
des Karlsruher Instituts für Technologie (KIT) genehmigte Dissertation

von MSc. Yoshiyuki Sakai aus Fukuoka, Japan

Tag der mündlichen Prüfung: 3. Juni 2016
Referent: Prof. Dr. Markus Uhlmann
Korreferent: Prof. Dr. Jörg Schumacher

Impressum

KIT Scientific Publishing

Karlsruher Institut für Technologie (KIT)
KIT Scientific Publishing
Straße am Forum 2
D-76131 Karlsruhe

KIT Scientific Publishing is a registered trademark
of Karlsruhe Institute of Technology.
Reprint using the book cover is not allowed.

www.ksp.kit.edu

Print on Demand 2019 – Gedruckt auf FSC-zertifiziertem Papier

ISSN 1439-4111
ISBN 978-3-7315-0790-1
DOI 10.5445/KSP/1000082066

"Always look on the bright side of life"

—Eric Idle

To the memory of Mineko
(1947–2016)

Abstract

Fluid flow through a straight duct with a rectangular cross-section exhibits turbulence-induced secondary motions of small amplitude, but with large consequences for momentum, heat and mass transport. The corresponding open duct flow (featuring a free surface) is characterised by a distinct secondary flow pattern, and is of great engineering interests such as man-made canals and urban rivers. Despite its high technical relevance, the detailed mechanism of the open duct secondary flow lacks our fundamental understanding, partly due to the required high computational cost to simulate the flow with adequate spatial and temporal resolutions. Consequently, the dependence of the mean secondary flow statistics on the flow Reynolds number and the geometrical aspect ratio is still to be addressed, together with the details of so-called "dip-phenomenon": the maximum of the average streamwise velocity is not found on the fluid surface but somewhat below.

The main aim of this study is therefore to address those fundamental questions based on the fully-resolved open duct turbulence dataset, generated by a series of direct numerical simulations over a range of Reynolds number and aspect ratio that are sufficiently broad.

The numerical approach employed here is based on a Fourier/Chevyshev pseudo-spectral technique, which had been thoroughly validated in prior to this study against both numerical and experimental benchmarks of the square closed duct turbulence (cf. Uhlmann et al., "*Marginally turbulent flow in a square duct*", J. Fluid Mech., **588**:153-162 (2007)). The rigid free-slip plane approximation was adopted to represent the free surface, which implies that the flow is on the zero Froude and Weber number limit. In total 38 open duct, and 10 additional closed duct simulations were performed over the course of the current study, with the bulk Reynolds number (Re_b) ranging from the marginal to fully-developed regime up to 7000, with the duct cross-sectional aspect ratio (A) from 0.5 to 8.

By analysing the generated dataset, we observed that, as in the closed duct counterpart, the buffer-layer coherent structures play a key role for the secondary flow formation in the open duct turbulence. Temporal tracking of those structures as well as an order-of-magnitude analysis showed that the unique open duct secondary flow pattern is a consequence of the vortex dynamics induced by the interactions between the quasi-streamwise vortices and the free surface.

The minimal sustainable Reynolds number of three narrow aspect ratios ($A = 0.5$, 1 and 2) were systematically determined, and it was observed that the minimal Reynolds number depends on aspect ratio again as in the closed duct counterpart.

Unlike the marginal open duct turbulence with $A = 1$, the flow in the cases of $A = 0.5$ and $A = 2$ exhibits drastic change in the mean secondary flow pattern when the flow Reynolds number is sufficiently close to the minimal limits. The appearance of such unique marginal flow patters was shown to be a consequence of some of the wall boundaries being too short to host a minimum set of the near-wall turbulent structures.
The Reynolds number dependence of the cross-sectional averaged mean secondary flow intensity was also investigated for $A = 1$. It was found that whilst the averaged intensity in the open duct flow scales with the bulk velocity (u_b), the same quantity in the closed duct flow scales with a mixed velocity scale ($\sqrt{u_\tau^3/u_b}$), where u_τ is the friction velocity based on the wall stress averaged over all no-slip boundaries.
Subsequently, the secondary flow intensity integrated only in the free-surface-normal direction was analysed. The integrated quantity, which is a function of the distance from the closest side-wall, was found to decay exponentially as the wall-distance increases. Furthermore, the intensity was found to become independent of aspect ratio if we further integrate it in the spanwise direction only over certain distance from the duct side-walls. Those results confirm the earlier observations that the secondary motions are confined in the vicinity of the side-walls even when the duct aspect ratio is large.
Finally, the properties of the velocity dip-phenomenon were investigated. It was found that the occurrence of the mean dip-phenomenon can be explained by the existence of large-scale low-velocity streaks in the duct mid-span region directly below the free surface. The corresponding submerged distance was found to be associated with the cross-sectional dimension of those streaks. Moreover, the critical aspect ratio to observe the dip-phenomenon over the entire duct span was found to be Reynolds number dependent in the moderate Reynolds number regime.

Zusammenfassung

Die Strömung durch einen geraden Kanal mit rechteckigem Querschnitt weist turbulenzinduzierte Sekundärbewegung mit kleiner Amplitude, aber großen Auswirkungen auf Impuls-, Wärme und Massentransport auf. Die dazugehörige offene Kanalströmung mit freier Oberfläche wird durch ein eindeutiges Sekundärströmungsmuster geprägt und ist von großer ingenieurstechnischer Bedeutung für beispielsweise künstlich angelegte Kanäle und urbane Flüsse.

Die mittlere Sekundärströmung des offenen Kanals besteht aus einem Paar kleinskaliger innerer Sekundärwirbel (*inner secondary vortices*), welches innerhalb der Ecken mit gemischten Randbedingungen angesiedelt ist, einem großskaligen äußeren Sekundärwirbelpaar (*outer secondary vortex*), welches sich unterhalb der freien Oberfläche befindet, sowie bodennahen Wirbeln ähnlich denen, die in der geschlossenen Kanalströmung beobachtet werden können. Trotz der hohen technischen Relevanz besteht eine merkliche Wissenslücke in Bezug auf den detaillierten Entstehungsmechanismus dieser Sekundärströmung im offenen Kanal.

Die Abhängigkeit des Phänomens sowohl von der Reynoldszahl, als auch dem Seitenverhältnis werden untersucht, zusammen mit Details des sogenannten *dip-phenomenon*, welches die Tatsache beschreibt, dass sich das Maximum der strömungsgerichteten Geschwindigkeitskomponente nicht an der Kanaloberfläche befindet.

Das Hauptziel dieser Arbeit ist es daher, die turbulente Kanalströmung mit Hilfe einer voll aufgelösten pseudo-spektralen direkten numerischen Simulation für eine Bandbreite an Reynoldszahlen und Seitenverhältnissen zu untersuchen. Dabei wurde die freie Oberfläche als schlupffrei und nicht deformierbar angenommen. Dies impliziert, dass die Froude- und die Weberzahl der Strömung null sind.

Ingesamt wurden im Rahmen dieser Arbeit 38 offene und zusätzlich zehn geschlossene Kanalsimulationen vom marginal turbulenten Regime bis hin zu Reynoldszahlen von $Re_b = 7000$ mit Seitenverhältnissen zwischen $A = 0.5$ und $A = 8$ durchgeführt. Die genannten Parameter wurden im Hinblick auf folgende Fragestellungen gewählt, um existierende Wissenslücken im Bereich der offenen Kanalströmungen zu füllen:

1. Identifizierung wesentlicher kohärenter Strukturen und Analyse derer Dynamik, welche für die Sekundärbewegungen im offenen Kanal verantwortlich ist;

2. Existenz von eindeutigen Strömungsmerkmalen im marginal turbulenten Bereich, entsprechend dem vierwirbeligen Strömungszustand im geschlossenen Kanal;

3. Abhängigkeit der mittleren Sekundärströmungsstrukturen bei voll ausgebildeter Turbulenz von der Reynoldszahl und dem Seitenverhältnis;

4. Abhängigkeit des *dip-phenomenon* von der Reynoldszahl und dem Seitenverhältnis;
5. Qualitative und quantitative Ähnlichkeiten und Unterschiede zur Sekundärbewegung in geschlossenen Kanälen;

Im Folgenden werden die Ergebnisse in Bezug auf die zuvor erwähnten Wissenslücken einzeln diskutiert mit Ausnahme von (5), auf welches im Verlauf eingegangen wird.

Wesentliche kohärente Strukturen und deren Dynamik Mittels ausgiebiger Wirbelbildungsanalyse wurden die quasi-strömungsgerichteten Wirbel, welche vorzugsweise in den Ecken mit gemischten Randbedingungen auftreten und dabei in Richtung dieser rotieren, als jene kohärente Strukturen identifiziert, welche wesentlich zur Bildung der mittleren inneren Sekundärströmung beitragen. Des Weiteren konnte festgestellt werden, dass jene quasi-strömungsgerichteten Wirbel unterhalb der freien Oberfläche, welche entgegen der Richtung der zuvor genannten Wirbel rotieren, zur mittleren äußeren Sekundärbewegung beitragen. Die charakteristische Wahrscheinlichkeitsverteilung der Wirbel ist mutmaßlich eine Konsequenz des Wirbelsortierungsmechanismus, welcher nur in der unmittelbaren Nähe der freien Oberfläche wirkt. Dieser transportiert die quasiströmungsgerichteten Wirbel in Spannweitenrichtung, wobei der Richtungssinn von der Rotationsrichtung abhängig ist. Eine zeitliche Verfolgung der Trajektorien relevanter Wirbel innerhalb einer repräsentativen offenen Kanalströmung quantifiziert die Wirbeldynamik und unterstützt die zuvor beschriebene Wirbelsortierungshypothese.

Offene Kanalströmung im marginal turbulenten Bereich bei verschiedenen Seitenverhältnissen Die Untersuchung legte offen, dass die minimal tragbare Reynoldszahl für Turbulenz im offenen Kanal, analog zur geschlossenen Kanalströmung, eine Funktion des Seitenverhältnisses ist. Minimale Reynoldszahlen wurden bestimmt für die drei Seitenverhältnisse $A = 0.5$, $A = 1$, sowie $A = 2$.
Im Vergleich zur marginal turbulenten Strömung im offenen Kanal mit $A = 1$, weist die Strömung im Fall der Seitenzahlverhältnisse $A = 0.5$ und $A = 2$ eine drastische Veränderung in Bezug auf das mittleren Sekundärströmungsmuster auf, wenn sich die Reynoldszahl in ausreichender Nähe zum zuvor beschriebenen Grenzwert befindet. Das Auftreten dieser einzigartigen marginalen Strömungsmuster erweist sich als Konsequenz aus der Tatsache, dass manche der Randkantenlängen zu kurz sind, um einen minimalen Satz wandnaher turbulenter Strukturen zu beheimaten.
Für die Seitenverhältnisse $A = 0.5$ und $A = 1$ zeigte die offene Kanalströmung, im Gegensatz zur geschlossenen, keinerlei zeitliche Variation der Orientierung. Jedoch zeigte die marginal turbulente Strömung im Fall $A = 2$ die Existenz einer zeitlich alternierenden Orientierung an, was weiterer Untersuchungen bedarf.

Reynoldszahlabhängigkeit der voll ausgeprägten turbulenten Kanalströmung Im Hinblick auf das mittlere Sekundärströmungsmuster weisen Größe und Position des mittleren inneren Sekundärwirbels eine starke Reynoldszahlabhängigkeit auf. Mit steigender Reynoldszahl wird dieser kontinuierlich kleiner und es findet eine Verschiebung in Richtung der Ecken mit

gemischten Randbedingungen statt. Die Position des mittleren strömungsgerichteten Wirbelstärkemaximums in Spannweitenrichtung zeigt hingegen keinerlei Reynoldszahlabhängigkeit. Derart einzigartige Skalierungseigenschaften in der offenen Kanalströmung konnten mit Hilfe der zuvor erwähnten Wirbelsortierungshypothese erläutert werden.
Die zuvor erwähnten Beobachtungen sind nur in der offenen Kanalströmung vorzufinden. Für die geschlossene Kanalströmung zeigte das Sekundärströmungsmuster keinerlei Abhängigkeit von der Reynoldszahl, wohingegen das strömungsgerichtete Wirbelstärkemuster von der Reynoldszahl abhängig ist.
Des Weiteren wurden die Skalierungseigenschaften der gemittelten mittleren Sekundärströmungsintensität im Querschnitt untersucht. Es wurde herausgefunden, dass die mittlere Intensität in der offenen Kanalströmung mit der *bulk*-Geschwindigkeit (u_b) skaliert, während dieselbe Größe in der geschlossenen Kanalströmung eine Proportionalität zu einer gemischten Geschwindigkeitsgröße ($\sqrt{u_\tau^3/u_b}$) aufweist. Hierbei ist u_τ die Schubspannungsgeschwindigkeit basierend auf der Wandschubspannung gemittelt über alle Wände mit Haftbedingung.

Seitenverhältnisabhängigkeit der voll ausgeprägten turbulenten Kanalströmung Frühere experimentelle und numerische Beobachtungen, welche nahelegten, dass die Sekundärströmung in der Nähe der Seitenwände auch für große Seitenverhältnisse begrenzt wird, konnten bestätigt werden. Infolgedessen zeigte die über die Kanaltiefe gemittelte Sekundärströmungsintensität keine Abhängigkeit vom Seitenverhältnis, falls sie über kleine Abstände von den Kanalseitenwänden in Spannweitenrichtung integriert wird. Zudem wurde für die offene und die geschlossene Kanalströmung festgestellt, dass die kanaltiefengemittelte Intensität mit zunehmendem Abstand von der nächsten Seitenwand, exponentiell abklingt. Die Abklingrate nähert sich hierbei für große Seitenverhältnisse asymptotisch einem konstanten Wert an.

Reynoldszahl- und Seitenverhältnisabhängigkeit des *dip-phenomenon* Das Auftreten des mittleren *dip-phenomenon* kann (im statistischen Sinne) mit der Existenz großskaliger *low-velocity streaks* in der Kanalmitte direkt unterhalb der freien Oberfläche erklärt werden. Die zugehörige Eintauchtiefe konnte in Zusammenhang mit der Querschnittsdimension jener *streaks* gebracht werden. Des Weiteren wurde zum ersten Mal die Abhängigkeit des zum Auftreten des *dip-phenomenon* im gesamten Strömungsbereich notwendigen, kritischen Seitenverhältnisses von der Reynoldszahl festgestellt. Die Resultate dieser Arbeit zeigen, dass das *dip-phenomenon* im Fall einer Strömung mit niedriger Reynoldszahl verschwindet, auch dann, wenn das Seitenverhältnis kleiner als der weithin akzeptierte kritische Wert von $A_{\text{crit}} \approx 2.5$ ist.

Acknowledgements

The research reported in this thesis was supported by the German Research Foundation (DFG) under the award code: UH242/3-1, which is gratefully acknowledged by the author. The simulations were performed partially at High Performance Computing Center Stuttgart (HLRS), and at the Steinbuch Center for Computing (SCC) of Karlsruhe Institute of Technology. The computer resources, technical expertise and assistance provided by these centres are gratefully acknowledged.
I would like to express my deep gratitude to Prof. Markus Uhlmann, my research supervisor and advisor, for his patient guidance, kind support and valuable critiques based on his vast knowledge in fluid dynamics during this research work. Without his selfless support and enthusiasm in science, I would have not been able to carry out my research and finish this thesis work in the present form. I would also like to thank Prof. Genta Kawahara, my external research advisor and our sincere research collaborator, for introducing myself to Prof. Uhlmann in the first place, and his hospitality and insightful advice during my visits at Osaka University in Japan.
I would like to sincerely thank Prof. Jörg Schumacher for accepting his role to be a member of my PhD committee, and for his disposition as the second referee of this thesis. I would like to thank also Prof. Peter Betsch, Prof. Robert Stieglitz and Prof. Martin Gabi for their interest in my research and disposition as members of my PhD committee.
I am particularly grateful to my current and former teammates: Todor, Agathe, Aman, Marco, Kad, Tiago, Chris and Michael, for the countless inspirations and enjoyments we shared. My experience to spend this dense time with such a great bunch of fluid enthusiasts was invaluable: Thanks guys. I would also like to extend my thanks to the other colleagues at IfH for their technical and administrative support, and particularly for the warm working atmosphere they created in the institute, in particular to: Helmut, Angelika, Heidi and Conny. I also owe many thanks to Ms. Diana Gudu for her long-suffering help in correcting my grammatical mistakes initially embedded all over this manuscript. Tens of lengthy unclear sentences, which I tend to produce so many during my writing, were detected and corrected by her careful eyes: Thanks D-chan.
Last but not least, I wish to thank my dear family in Japan, Hisae (mother), Yoshio (father) and Chiaki (sister) to list a few, for their love and understanding for allowing me to persuade my rather selfish passion in science, especially over this long period of my life being so far away from them. Without them, my life is meaningless. ARIGATOU.

Yoshiyuki Sakai
Karlsruhe, April 2016

Related Publications

The content of this thesis is based in parts on the following references, published during the course of this PhD thesis. The contributions of the co-authors are highly acknowledged.

Peer-reviewed publications

1. Y. Sakai and M. Uhlmann. High-resolution numerical analysis of turbulent flow in straight ducts with rectangular cross-section. In *High Perform. Comput. Sci. Eng. '15*. Springer International Publishing, 2016.

Conference contributions

1. Y. Sakai. High-resolution Numerical Analysis of Turbulent Flow in Straight Ducts with Rectangular Cross-section, Seventh European Postgraduate Fluid Dynamics Conference, Ilmenau (Germany), July 2014.

2. Y. Sakai, M. Uhlmann and G. Kawahara. High-resolution Numerical Analysis of Turbulent Flow in Straight Ducts with Rectangular Cross-section, 15th European Turbulence Conference, Delft (The Netherlands), August 2015.

Contents

List of Figures

List of Tables

Nomenclature

Mathematical symbols

$*^+$	quantity $*$ normalised by wall units (i.e. u_τ, δ_v)
$*'$	fluctuation of a quantity $*$ about its mean $\langle * \rangle$
$\hat{*}$	field value $*$ in Fourier-physical-physical space in x, y, z-directions
$\tilde{*}$	field value $*$ in Fourier-Chebyshev-Chebyshev space
$\langle \cdot \rangle$	averaging operator in time and in all homogeneous directions
$\langle \cdot \rangle_{\mathrm{cs}}$	averaging operator only within volume of educed coherent structures
$\langle \cdot \rangle_{\mathrm{inst}}$	averaging operator in numerous instantaneous snapshots
$\langle \cdot \rangle_{\mathrm{rms}}$	root-mean-square operator
$\langle \cdot \rangle_{\mathrm{V}}$	averaging operator in numerical domain volume
$\max\{\cdot\}$	maximum operator
$\mathcal{O}(\cdot)$	order-of-magnitude
∇	Nabla operator, $[\partial_x, \partial_y, \partial_z]^T$
∇^2	Laplace operator, $[\partial_x^2, \partial_y^2, \partial_z^2]^T$
$\nabla_\perp^2$	cross-sectional second derivative operator, $(\partial_y^2 + \partial_z^2)$
d/di	derivative in i-direction
$\partial/\partial i$, ∂_i	partial derivative in i-direction

Lower-case Roman symbols

a	side-wall distance coefficient
c	Helmholtz coefficient
d	distance between a vortex and its mirrored image over free-slip plane
f	friction factor
$\hat{f}$	right-hand-side of Poisson/Helmholtz problem
h	channel (closed duct) half-height
k	turbulent kinetic energy
l	characteristic vortex length-scale
$l_{\mathscr{A}}$	cross-sectional length-scale of educed velocity streaks
$l_{\mathscr{A}\max}$	maximum cross-sectional length-scale of educed velocity streaks
l_{DI}	upper limit of dissipation range
l_{EI}	upper limit of universal equilibrium range

l_x, l_y, l_z	length-scales of self-similar vortex cluster
l_0	the largest length-scale of vortical motions
p	pressure
$\boldsymbol{r}_{i,j}$	position difference vector between vortex i and j
r_{th}	position difference threshold for vortex tracking
t	time
t^n	discrete time at temporal step n
$\boldsymbol{u}$	velocity vector, $[u_1, u_2, u_3]^T = [u, v, w]^T$
u, v, w	velocity vector component in x, y, z-direction
$\boldsymbol{u}^*$	predicted non-divergence-free velocity in fractional step
u_b	bulk velocity
$([u_{\mathrm{v}}^m, v_{\mathrm{v}}^m, w_{\mathrm{v}}^m])_l$	velocity of vortex trajectory l at discrete time m
u_{th}	u' threshold value for cross-sectional streak eduction
$u_{\mathrm{w}}, v_{\mathrm{w}}, w_{\mathrm{w}}$	velocity components at no-slip walls
u_η	Kolmogorov velocity-scale
u_τ	friction velocity
$\boldsymbol{x}$	position vector, $[x_1, x_2, x_3]^T = [x, y, z]^T$
x, y, z	Cartesian coordinates (streamwise, wall-normal, spanwise directions)
$(x_{\mathrm{c}}, y_{\mathrm{c}}, z_{\mathrm{c}})_i$	centroid coordinates of educed vortex i
$(\tilde{x}_{\mathrm{c}}, \tilde{y}_{\mathrm{c}}, \tilde{z}_{\mathrm{c}})_i$	shifted centroid coordinates of educed vortex i
$([x_{\mathrm{v}}^m, y_{\mathrm{v}}^m, z_{\mathrm{v}}^m])_l$	centroid mid-point coordinates associated with $([u_{\mathrm{v}}^m, v_{\mathrm{v}}^m, w_{\mathrm{v}}^m])_l$
$y_{\mathrm{c}}, z_{\mathrm{c}}$	centroid coordinates of educed cross-sectional velocity streaks
$y_{\max}$	y-coordinate of maximum $\langle u \rangle$
z''	distance from the closest side-wall in z-direction

Upper-case Roman symbols

$\mathscr{A}$	duct cross-sectional area
A	duct aspect ratio
A_{crit}	critical aspect ratio for global dip-phenomenon
$\mathscr{A}_{\mathrm{s}}$	cross-sectional area of educed velocity streaks
B	logarithmic constant in log-law
D	discriminant of cross-sectional velocity gradient tensor
D_{H}	duct hydraulic diameter
$\boldsymbol{D}_{l,m}$	first-order cross-sectional derivative matrix
$\boldsymbol{D}_i^2$	second-order cross-sectional derivative matrix in i-direction
$\boldsymbol{F}$	right-hand-side matrix of Poisson/Helmholtz problem
H	open channel (duct) full-height
I	indicator function based on Ω_i
$\check{I}$	indicator function based on $\check{\Omega}_i$
$\mathscr{L}$	characteristic length-scale
L_x, L_y, L_z	numerical domain size in x, y, z-directions

N_s	number of educed velocity streaks
N_v	number of educed vortices
$\mathscr{P}$	production term of TKE budget equation
P	duct wetted perimeter
$\boldsymbol{P}$	eigenvalue matrix of $\boldsymbol{D}^2$
P^*	pressure gradient computed for constant mass-flow rate
Q	prescribed mass-flow rate
Re	Reynolds number
Re_b	bulk Reynolds number based on u_b and $h(H)$
Re_{crit}	critical Reynolds number
Re_l	Reynolds number based on length-scale l
Re_0	Reynolds number based on U_0 and δ
Re_η	Reynolds number based on u_η and η
Re_τ	friction Reynolds number based on u_τ and δ_v
$S_{\text{est}}, V_{\text{est}}$	estimated surface area/volume of typical buffer-layer vortices
S'_{ij}	fluctuation part of rate-of-strain tensor
S_i	squared ω_x averaged in x-direction integrated over Ω_i
$\check{S}_i$	squared ω_x averaged in x-direction integrated over $\check{\Omega}_i$
$S_{\text{th}}, V_{\text{th}}$	area/volume threshold values to filter educed vortices
T_l	age of vortex along trajectory l
$\tilde{T}_l$	life-time of vortex trajectory l
T_n	Chebyshev polynomial of degree n
T_s	overall time-interval of snapshot dataset
$\mathscr{U}$	characteristic velocity-scale
$\boldsymbol{U}$	cross-sectional velocity matrix
U_{rms}	cross-sectional averaged $\langle u' \rangle_{\text{rms}}$
U_0	free-stream (centre-line) velocity
$U_\perp$	cross-sectional averaged secondary flow intensity
$U_\perp^*$	secondary flow intensity averaged only over some distance from side-wall
$U_{\perp\text{rms}}$	cross-sectional averaged $\langle (v' + w') \rangle_{\text{rms}}$
$U_{\sqrt{\langle v'w' \rangle}}$	cross-sectional averaged $\sqrt{\langle v'w' \rangle}$
V_{box}	volume of numerical domain
V_c	cubical volume used to compute $V_{i \wedge j}$
V_i	volume of educed vortex i
V_{ind}	velocity-scale induced by mirrored-dipole mechanism
$V_{i \wedge j}$	overlapping volume of vortex i and j
V_{turb}	turbulent velocity-scale near free-slip plane

Greek symbols

α	dip-correlation factor
$\alpha_k, \gamma_k, \zeta_k$	Runge-Kutta coefficients at kth step

α_{r}	user-specified distance coefficient for vortex tracking
α_{v}	user-specified overlapping volume-ratio coefficient for vortex tracking
γ	circulation of vortex
δ	boundary-layer thickness
δ_v	viscous length-scale
Δt	time-step size of temporal integration
Δt_{s}	time-interval between subsequent snapshots
$([\Delta x_{\mathrm{v}}^m, \Delta y_{\mathrm{v}}^m, \Delta z_{\mathrm{v}}^m])_l$	spatial displacement along vortex trajectory l
ε	dissipation
$\tilde{\varepsilon}$	pseudo-dissipation
η	Kolmogorov length-scale
κ	von Kárman constant
$\mathbf{\Lambda}$	eigenvector matrix of $\boldsymbol{D}^2$
ν	fluid kinematic viscosity
ρ	fluid density
σ_u	volume-averaged $\langle u' \rangle_{\mathrm{rms}}$
τ_{B}	typical time-interval of bursting events
τ_{w}	wall stress
τ_η	Kolmogorov temporal scale
ϕ^k	pseudo-pressure at kth Runge-Kutta step
ϕ^*	duct shape factor
ψ	cross-sectional streamfunction
$\boldsymbol{\omega}$	vorticity vector
Ω	enstrophy averaged over cross-section
ω_i	vorticity in i-direction
Ω_i	ith sub-domain of duct cross-section split into 4 equiareal sections
$\check{\Omega}_i$	ith sub-domain of duct cross-section split into 3 equiareal sections
$\omega_{x,\mathrm{th}}$	ω_x threshold value used to educe vortices

Abbreviations

CFL	Courant-Friedrich-Lenny
CGL	Chebyshev-Gauss-Lobatto
DNS	direct numerical simulation
FCT	fast cosine transform
FFT	fast Fourier transform
RANS	Reynolds-averaged Navier-Stokes
TKE	turbulent kinetic energy

Chapter 1

Introduction

1.1 Background and motivations

The study of wall-bounded flow has tremendous impact upon our daily life. To highlight the importance, it is useful to list few examples from the broad range of relevant physical phenomena in our surrounding nature, for example from the large-scale end: Atmospheric boundary layer of our planet Earth, and the oceanic currents inside the large-scale bays like Gulf of Mexico; and from the smaller scale spectrum: Stream of blood that circulating our bodies and the stream of air that we unconsciously induce into our respiratory system, and the list goes on. Moreover, a vast number of engineering applications that unavoidably contain wall-bounded flow should not be forgotten, such as intercontinental network of pipelines transporting natural gas from Russia to Europe, internal-combustion engines of the cars and ships that transport our everyday essentials; the cooling systems of those engines that are required to operate the power train efficiently and reliably. In fact, the economical impact of such kind of flow is so large that approximately one-quarter of the entire industrial and commercial energy consumption is spent by the wall-bounded flow, through pipes, ducts and vehicles that move through air and water (Jiménez (2013)).

Wall-bounded flow is also of particular interests to the engineers in the civil engineering discipline, in a form of, for instance, rivers, man-made canals and domestic sewage systems. In many cases, such civil-engineering-related flow is some sort of *duct flow* that is a stream of fluid transported through ducts with rectangular/trapezoidal cross-section, enclosed by four solid walls forming four sharp corners (cf. Figure 1.1(a)). Moreover, the duct flow often incorporates a free surface, such as the water-air interface of the urban canals, consequently forming mixed-boundary corners (cf. Figure 1.1(b)). Henceforth we shall refer to duct flow of the former type as *closed duct flow*, whereas the latter type with free surface will be referred to as *open duct flow* as a convention.

Both kinds of flow are known to exhibit weak secondary motions (or *secondary flow*) in their cross-sectional directions when the state of such flow becomes *turbulent* (more explanations on the term *turbulent flow* or simply *turbulence* will be provided in the following section). Appearance of the well-defined vortical patterns of the turbulent secondary flow, which are presented in Figure 1.2, is a feature of the mean flow that should not be interpreted as the permanent existence of such large-scale streamwise vortices.

Figure 1.1: Practical examples of (a) closed, and (b) open duct flow: (a) air ventilation duct; (b) Magdeburg Water Bridge in Magdeburg, Germany.

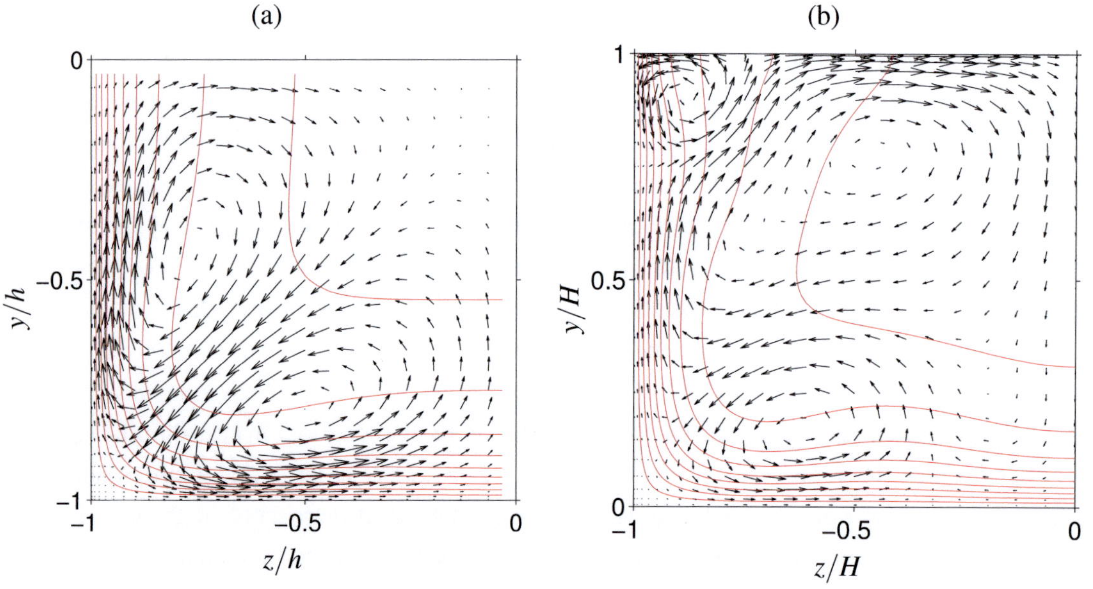

Figure 1.2: Mean secondary flow vector and mean streamwise velocity contour of: (a) closed and (b) open duct flow at $Re_b = 2205$. Only lower left quadrant and left half of the duct cross planes are shown respectively.

Despite its high technical and physical importance, and therefore long history of research on the topic, the detailed mechanisms of the turbulent duct secondary flow remain largely unknown chiefly due to the weak intensity (few percent of the flow bulk velocity) which makes it particularly challenging to analyse quantitatively in experiments and direct numerical simulations. Similarly, due to the strong link between the generation mechanism of the turbulent secondary flow and anisotropy of the turbulent velocity fluctuations, many of the common turbulence models fail to reproduce the secondary flow even qualitatively.

The last few decades of explosive growth of computational power, however, succeeded to add a new perspective in the above picture: A rise of the modern two/three-dimensional experimental measurement techniques, and more importantly for the current work, *direct numerical simulations* (or *DNS* in short) became available in the fundamental fluid dynamics research. In return to the high computational effort, the direct numerical simulation technique is able to resolve every spatial and temporal scale of fluid motions without any aid of the turbulence modelling, including the anisotropy of the near-wall turbulence that is in fact crucial to study any kind of wall-bounded turbulent flow. Consequently, the numerical study of the square closed duct turbulence by Gavrilakis (1992) set an important milestone in the field by being the first fully-resolved direct numerical simulation of its kind. Subsequently, our understanding of the closed duct secondary flow has been significantly advanced, though it is still far from being conclusive, including the importance of *turbulent coherent structures* demonstrated probabilistically by Uhlmann et al. (2007) and Pinelli et al. (2010) with their square closed duct pseudo-spectral DNS. Note that the precise definitions of the term "*turbulent coherent structures*", or alternatively "*coherent motions in turbulent flow*", will be introduced shortly.

On the contrary, our understanding of the corresponding open duct secondary flow is still severely limited, owing largely to the fact that no fully-resolved open duct simulation has been performed to date. It is therefore our main motivation in the current work to perform a series of open duct direct numerical simulations in order to advance our understanding in the topic of open duct secondary flow to the current level of the closed duct counterpart and beyond. It is important to note that our special attention will be placed upon the importance of the turbulent coherent structures throughout this study.

Prior to the specification of more detailed scope and objectives of this work, some fundamental concepts necessary to hold any fruitful discussion upon the wall-bounded turbulence need to be introduced in the following sections. Those conceptual introduction will be followed by the literature surveys conducted for this thesis work.

1.2 Fundamentals of turbulence

Most of natural and engineering flow phenomena are in a complex, chaotic and therefore in detail unpredictable state called *turbulence*, in contrast to the accompanying state called *laminar* flow that is simple and in principle predictable to arbitrary accuracy. The fundamental distinction between those states was first made by Osborne Reynolds in his historic experimental works with a pipe flow set-up equipped with a thin stream of dye injected into the flow (cf. Reynolds (1883), Reynolds (1895)). He discovered that the motions of fluid elements visualised by the injected dye exhibit a transition from quiescent and well-organised laminar state to more dynamic and disorganised turbulent state through varying fundamental flow parameters, such as flow velocity, pipe diameter and viscosity of the working fluid (cf. Figure 1.3).

Moreover, he realised that the transitional process can be characterised by a critical value of the single dimensionless number that bears his name today, *Reynolds number* (*Re*), which represents a ratio between the inertial and viscous forces acting upon fluid elements, viz.

$$Re = \frac{\mathscr{U}\mathscr{L}}{\nu} \tag{1.1}$$

where $\mathscr{U}$ and $\mathscr{L}$ are characteristic velocity- and length-scales (the mean flow velocity and the pipe diameter for his experiments), whereas ν represents kinematic viscosity of the flow. Considering

the fact that in a Newtonian fluid the viscous stresses are proportional to the the spatial gradients of the velocity field, and act as a "damper" or a "stabiliser" of fluid motions by converting kinetic energy into heat ("*viscous dissipation*", or simply "*dissipation*"), the observed Reynolds number dependency can be qualitatively explained as follows.
In case of viscosity-dominant sub-critical Reynolds number flow (i.e. $Re \ll Re_{\text{crit}}$, where Re_{crit} is a critical Reynolds number), the flow stays laminar since any fluctuation of motions caused by disturbances from the surrounding environment (e.g. slight temperature variations, acoustic noises, minute roughness elements on the boundaries) are quickly dissipated into heat and therefore cannot be sustained. Conversely in case of supercritical flow ($Re \geq Re_{\text{crit}}$), the injected energy through those disturbances cannot be dissipated at once, so the physical system of fluid flow is somehow forced to transform its behaviour to compensate the insufficiency of energy dissipation, namely by increasing the velocity gradient introduced by localised circulating motions.
Such distinguishable geometries of coherent motions are often referred to as *vortices* or *eddies*, and the geometrical complexity of the flow pattern is known to increase as the flow Reynolds number also increases, in order to achieve higher degree of velocity gradient and therefore better energy dissipation efficiency. The evolution of the complexity is manifested by broadening the range of the length-scale of the vortical motions, which implies the co-existence of the *self-similar* vortices, which span over a range of length-scales depending on the flow Reynolds number. This aspect of the turbulent flow will be detailed in the following sections with some aid of the mathematical descriptions of the fluid motion.

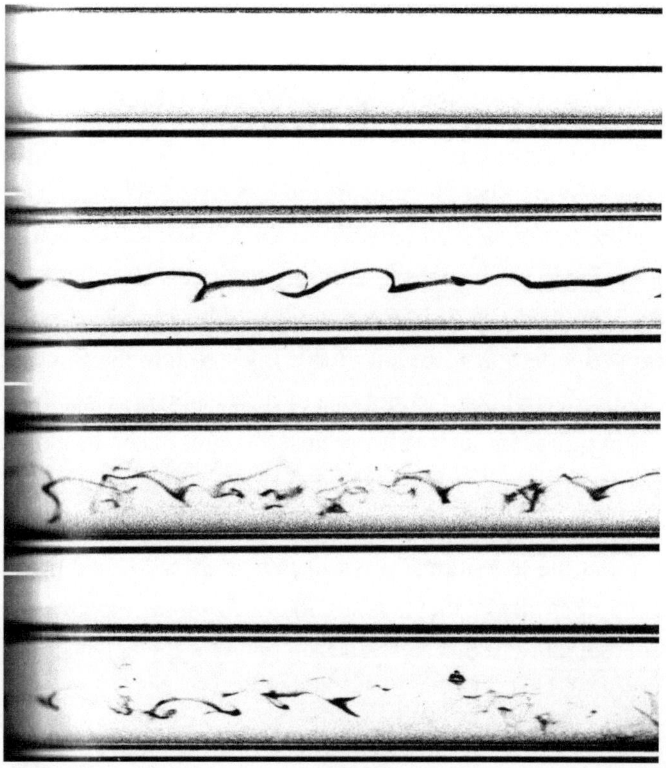

Figure 1.3: Reproduced Reynolds' pipe experiments visualised by injected dye. Reynolds number is increased from top to bottom. This figure is from Van Dyke (1982).

Details of the transition mechanisms to turbulence as well as the precise value of the critical Reynolds number are, on the other hand, out of the scope of this thesis work, therefore we limit ourselves to discuss only briefly at this point. The transition mechanisms can be categorised into two groups: the supercritical and sub-critical transitions, where the former can be triggered by infinitesimal disturbances and the continuous amplification of those disturbances leads to a complete transition; whilst the latter requires finite disturbances and exhibits locally stable states. The locally stable states manifest as a variety of forms depending on the different wall-bounded flow configurations, such as the spatially localised turbulent *spots* and *bands* observed in the plane channel and wide duct flow; whereas *puffs* and *slugs*, which occupy the entire cross-section but isolated in the streamwise direction can be observed in the pipe and narrow duct flow (cf. Figure 1.4). Note for the study of the subcritical transition, a significant progress has been made recently by a number different groups for various canonical wall-bounded flow configurations (e.g. Hof et al. (2006), Barkley et al. (2015), Sano and Tamai (2016)). The interested readers are encouraged to refer to those references as well as the standard textbooks of the subject (e.g. Schmid and Henningson (2001)).

In the following few sections, a set of fundamental mathematical concepts and tools that are necessary to describe the turbulent fluid motions both qualitatively and quantitatively will be introduced. To concentrate ourselves upon the concepts rather than the mathematical completeness, only the physical quantities that are defined on the Cartesian coordinate system will be considered. The physical consequences of the introduced mathematical concepts will be elaborated wherever necessary. Those sections will be followed by the literature survey of the wall-bounded turbulence as well as the duct turbulence and its secondary motions.

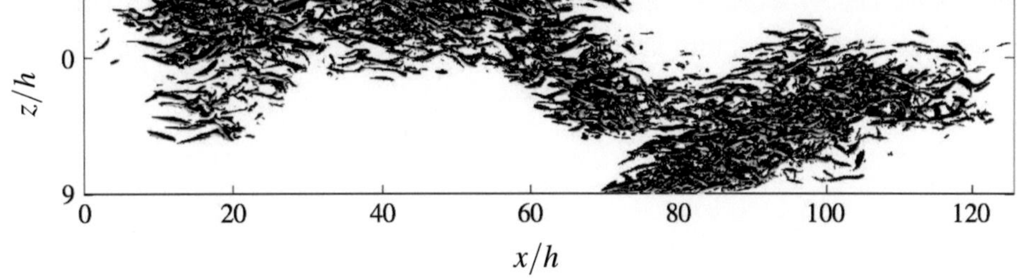

Figure 1.4: Examples of turbulent (a) *puff* and (b) *band*. The flow configurations are the rectangular closed duct with the aspect ratios at: (a) 2 and (b) 9. The visualised turbulent structures are the iso-contours based on the Q-criterion of Hunt et al. (1988). The figures are from Takeishi et al. (2015).

1.2.1 Governing equations of fluid motion

The chaotic nature of the turbulent flow arises from the non-linearity of the governing equations of fluid motions, so-called *Navier-Stokes equations*. If we consider a case of *incompressible* constant density flow, then the governing equations in the vector notation can be expressed as follows:

$$\nabla \cdot \boldsymbol{u} = 0 \,, \tag{1.2a}$$

$$\frac{\partial \boldsymbol{u}}{\partial t} + (\boldsymbol{u} \cdot \nabla)\boldsymbol{u} = -\frac{1}{\rho}\nabla p + \nu \nabla^2 \boldsymbol{u} \,, \tag{1.2b}$$

where p and $\boldsymbol{u}$ are the pressure scalar and velocity vector fields respectively, whereas ρ is the fluid density which is considered to be constant here. In fact, it is a common practice in the numerical investigations of the incompressible flow to set $\rho = 1$ in order to omit the quantity from the governing equations. Note that the introduced governing equations are nothing but a combination of *continuity* constraint (aka. *divergence-free* constraint), and the momentum balance in each direction. Note also that we will employ the aforementioned vector notation $\boldsymbol{u}$ and the corresponding index notation $u_i = [u, v, w]^T$ with $i = 1,2,3$ to express the three-dimensional velocity interchangeably hereafter. Moreover, it is the convention of numerical wall-bounded flow community that we normally set the primary *streamwise* direction with $i = 1$, whilst the directions with $i = 2$ and 3 are *wall-normal* (or *free-slip plane normal* in a case of open duct flow) and *spanwise* directions respectively. The corresponding spatial coordinate system are defined as $\boldsymbol{x} = [x_1, x_2, x_3]^T = [x, y, z]^T$. Consequently, Equation 1.2 can be expressed in the alternative index notation as:

$$\frac{\partial u_i}{\partial x_i} = 0 \,, \tag{1.3a}$$

$$\frac{\partial u_j}{\partial t} + \frac{\partial u_i u_j}{\partial x_j} = -\frac{1}{\rho}\frac{\partial p}{\partial x_j} + \nu \frac{\partial^2 u_j}{\partial x_i^2} \,. \tag{1.3b}$$

Note we employed the Einstein summation convention to express the summation over all i and j. Let us also introduce a key concept applied frequently in practice to deal with the random nature of the turbulent flow, namely *Reynolds decomposition*, which decomposes any spatially and temporally varying field value $\phi(\boldsymbol{x},t)$ into the ensemble-averaged value (or *mean*, denoted by $\langle \phi(\boldsymbol{x}) \rangle$), and the deviation from the mean (i.e. $\phi'(\boldsymbol{x},t)$), viz.

$$\phi(\boldsymbol{x},t) = \langle \phi(\boldsymbol{x}) \rangle + \phi'(\boldsymbol{x},t) \,. \tag{1.4}$$

Note in case of the flow with some statistically homogeneous directions, such as the streamwise direction in the duct flow, the mean value becomes a function of the remaining inhomogeneous spatial coordinates. Note also that the notation of the spatio-temporal dependency of those flow variables will be omitted for convenience hereafter (e.g. $\boldsymbol{u}(\boldsymbol{x},t) = \boldsymbol{u}$), unless otherwise stated. Some key mathematical relations between the decomposed quantities and the averaging operation denoted by $\langle \cdot \rangle$ should be mentioned, and they can be summarised as follows:

$$\langle\langle\phi\rangle\rangle = \langle\phi\rangle\,, \tag{1.5a}$$
$$\langle\phi'\rangle = 0\,, \tag{1.5b}$$
$$\langle\phi'\langle\phi\rangle\rangle = 0\,, \tag{1.5c}$$
$$\langle\phi'\phi'\rangle \neq 0\,. \tag{1.5d}$$

Consequently, a set of equations which describe the mean fluid motions, namely *Reynolds-averaged Navier-Stokes* equations (or *RANS* equations in short), can be derived by applying the Reynolds decomposition to all field quantities of the governing equations and organising the resulting equations by Equation 1.5, viz.

$$\frac{\partial\langle u_i\rangle}{\partial x_i} = 0\,, \tag{1.6a}$$
$$\frac{\partial\langle u_i\rangle}{\partial t} = \frac{\partial\langle u_i\rangle\langle u_j\rangle}{\partial x_j} + \frac{\partial\langle u_i'u_j'\rangle}{\partial x_j} = -\frac{1}{\rho}\frac{\partial\langle p\rangle}{\partial x_j} + \nu\frac{\partial^2\langle u_j\rangle}{\partial x_i^2}\,. \tag{1.6b}$$

The term $\langle u_i'u_j'\rangle$ is often referred to as *Reynolds stresses*, forming *Reynolds stress tensor* that is symmetric about the diagonal *normal stress* components where $i=j$. The off-diagonal components (i.e. $i\neq j$) are generally referred to as the *shear stress* components of the Reynolds stress tensor.

1.2.2 Turbulent kinetic energy and its transport equations

At this point, it is also convenient for the following discussions to introduce the notion of the kinetic energy of the velocity fluctuations, namely *turbulent kinetic energy* (or *TKE* in short), as

$$k \equiv \frac{1}{2}\langle \mathbf{u}'\cdot\mathbf{u}'\rangle = \frac{1}{2}\langle u_i'u_i'\rangle\,. \tag{1.7}$$

The corresponding transport equations can be derived from Equation 1.3 and 1.4, which can be expressed as:

$$\frac{\partial k}{\partial t} + \langle u_j\rangle\frac{\partial k}{\partial x_j} + \frac{1}{\rho}\frac{\partial\langle u_j'p'\rangle}{\partial x_j} + \frac{1}{2}\frac{\partial\langle u_i'u_i'u_j'\rangle}{\partial x_j} - 2\nu\frac{\partial\langle u_i'S_{ij}'\rangle}{\partial x_j} = -\langle u_i'u_j'\rangle\frac{\partial\langle u_i\rangle}{\partial x_j} - 2\nu\langle S_{ij}'S_{ij}'\rangle\,, \tag{1.8}$$

where S_{ij}' is the fluctuation part of the rate-of-strain tensor:

$$S_{ij}' \equiv \frac{1}{2}\left(\frac{\partial u_i'}{\partial x_j} + \frac{\partial u_j'}{\partial x_i}\right)\,. \tag{1.9}$$

Note the correlation term between the Reynolds stress and the mean velocity spatial gradient is generally positive in a case of the wall-bounded turbulence, and commonly referred to as *production* term $\mathscr{P}$ of the turbulent kinetic energy:

$$\mathscr{P} \equiv -\langle u_i'u_j'\rangle\frac{\partial\langle u_i\rangle}{\partial x_j}\,. \tag{1.10}$$

This terminology is associated with the common understanding that the kinetic energy of the mean flow is drained and transferred to the fluctuation energy field through this term.
Similarly, the term with viscosity is strictly positive due to the squared part, and referred to as *dissipation* term ε of the turbulent kinetic energy:

$$\varepsilon \equiv 2\nu \langle S'_{ij} S'_{ij} \rangle \tag{1.11}$$

Note the negative sign in front of the dissipation term in Equation 1.8, which indicates the term acts as sink. In addition, by introducing pseudo-dissipation $\tilde{\varepsilon}$ as:

$$\tilde{\varepsilon} \equiv \varepsilon - \nu \frac{\partial^2 \langle u'_i u'_j \rangle}{\partial x_i \partial x_j} , \tag{1.12}$$

Equation 1.8 can be then simplified as:

$$\frac{\partial k}{\partial t} + \langle u_j \rangle \frac{\partial k}{\partial x_j} + \frac{1}{\rho} \frac{\partial \langle u'_j p' \rangle}{\partial x_j} + \frac{1}{2} \frac{\partial \langle u'_i u'_i u'_j \rangle}{\partial x_j} - \nu \frac{\partial^2 k}{\partial x_j^2} = \mathscr{P} - \tilde{\varepsilon} . \tag{1.13}$$

Note the difference between ε and $\tilde{\varepsilon}$ is often negligible (Pope, 2000).

1.2.3 Vorticity and its transport equations

Since the turbulent flow dynamics is dominated by the vortical fluid motions, it is also necessary for many quantitative analysis to introduce the *vorticity* vector $\boldsymbol{\omega}$ of the fluid velocity fields, which can be defined as the curl of a velocity vector field, viz.

$$\boldsymbol{\omega} \equiv \nabla \times \boldsymbol{u} , \tag{1.14}$$

whose component-wise expressions are

$$[\omega_x, \omega_y, \omega_z]^T \equiv \left[\frac{\partial w}{\partial y} - \frac{\partial v}{\partial z}, \frac{\partial u}{\partial z} - \frac{\partial w}{\partial x}, \frac{\partial v}{\partial x} - \frac{\partial u}{\partial y} \right]^T . \tag{1.15}$$

The corresponding transport equations in the vector notation can be expressed as:

$$\frac{\partial \boldsymbol{\omega}}{\partial t} + (\boldsymbol{u} \cdot \nabla) \boldsymbol{\omega} = (\boldsymbol{\omega} \cdot \nabla) \boldsymbol{u} + \nu \nabla^2 \boldsymbol{\omega} . \tag{1.16}$$

which can be written equivalently in the following index notation:

$$\frac{\partial \omega_j}{\partial t} + u_i \frac{\partial \omega_j}{\partial x_i} = \omega_i \frac{\partial u_j}{\partial x_i} + \nu \frac{\partial^2 \omega_j}{\partial x_i^2} , \tag{1.17}$$

where $[\omega_1, \omega_2, \omega_3]^T = [\omega_x, \omega_y, \omega_z]^T$.

Finally we conclude with the transport equation of the mean streamwise vorticity which is frequently employed to analyse the mean secondary motion of duct turbulence (e.g. Demuren and Rodi (1984); Gavrilakis (1992)):

$$\frac{\partial \langle \omega_x \rangle}{\partial t} + \langle u_i \rangle \frac{\partial \langle \omega_x \rangle}{\partial x_i} =$$
$$\langle \omega_i \rangle \frac{\partial \langle u \rangle}{\partial x_i} + \frac{\partial}{\partial x} \left\{ \frac{\partial \langle u'v' \rangle}{\partial z} + \frac{\partial \langle u'w' \rangle}{\partial y} \right\} + \frac{\partial^2}{\partial y \partial z} \{ \langle v'v' \rangle - \langle w'w' \rangle \} + \left\{ \frac{\partial^2}{\partial z^2} - \frac{\partial^2}{\partial y^2} \right\} \langle v'w' \rangle$$
$$+ \nu \frac{\partial^2 \langle \omega_x \rangle}{\partial x_i^2} . \quad (1.18)$$

1.2.4 The scales of turbulent motions

The remaining part of this section will be devoted to introduce some mathematical and statistical concepts of the length-scales of turbulent motions. Most of the definitions and notations were adopted from the standard textbook of turbulence by Pope (2000), therefore interested readers are encouraged to refer to the Chapter 6 of the book for more details.

A fundamental aspect of the turbulent flow is the co-existence of the large-scale energy-containing eddies, whose size and geometry are under a large influence from the specificity of the flow domain, both geometry and boundary conditions, alongside a broad spectrum of the less energetic small-scale eddies with increasing universality of their physical properties. Such observation can be qualitatively explained based on the transport properties of the turbulent kinetic energy that was introduced in Section 1.2.2. As it was briefly mentioned above, the energy from (the spatial gradient of) the mean velocity field enters into the turbulence through the production mechanism shown in the transport equations (cf. Equation 1.13 with 1.12). Since the mean velocity profile is largely owed by the bulk geometry of the flow domain, the geometries of motions corresponding to the injected energy from the mean to fluctuating fields consequently have the large spatial and temporal scales as well as the anisotropy of those quantities inherited from their origin.

Consequently, those large-scale anisotropic vortices are unstable in nature, therefore they brake down into smaller vortices relatively soon. This cascading process of the vortex length-scale towards the smaller end continues successively until the length-scale becomes small enough to be affected by the viscous dissipation process, which eventually dissipates the kinematic energy into heat. This notion of step-by-step inter-scale energy transfer was first introduced by Richardson (1922), which is referred to as *forward energy cascade* today. Moreover, the hierarchical organisation of the length-scale of the vortical motions is evident from the well-known *self-similar* characteristics of the turbulent flow. It is also worth mentioning that the opposite sense of energy transfer can be observed in the real-world turbulence (i.e. small-to-large vortex coalescence), although it is much less frequent compared to the predominant large-to-small cascading. Such process is often referred to as *inverse energy cascade*.

More quantitative discussions of the hierarchical length-scale distribution became later available by the historic work of Kolmogorov (1941), by means of his three signature hypotheses known as *Kolmogorov hypotheses*, which will be introduced as following.

Kolmogorov considered specifically the very high-Reynolds number turbulent flow and its statistical properties, and divided the range of length-scale into two parts, namely *energy-containing range* and *universal equilibrium range*, separated by a representative scale l_{EI}. When a vortical length-scale of concern, say l, is larger than l_{EI} (i.e. $l > l_{EI}$, inside the energy-containing range), then the vortex contains a large portion of kinetic energy with a high degree of anisotropy and some specific features from the flow domain. Inside this range the characteristic length-scale $\mathscr{L}$ and the largest length-scale of the turbulent vortical motion l_0 reside. Conversely if l is smaller than l_{EI}, then the corresponding statistical properties hold a universal from which is determined uniquely by ν and ε (**Kolmogorov's first similarity hypothesis**).

Inside the universal equilibrium range, he introduced two additional classes of length-scale, namely *inertial subrange* and *dissipation range*. In case of the length-scale l is within the inertial subrange, then the statistical properties depend solely on the dissipation ε (**Kolmogorov's second similarity hypothesis**). If, on the other hand, l is within the dissipation range, then the viscosity effect is not negligible, therefore the statistical properties are functions of both ε and ν.

From the above dimensional consideration, Kolmogorov defined the theoretical lower limit of the eddy motion (η) and the corresponding velocity (u_η) and temporal (τ_η) scales relevant inside the dissipation range, which are referred to as *Kolmogorov scales* today. The definitions of those scales are designed to yield the following Reynolds number:

$$Re_\eta = \frac{u_\eta \eta}{\nu} \equiv 1 \,, \tag{1.19}$$

and the outcomes can be expressed as:

$$\eta \equiv \left(\frac{\nu^3}{\varepsilon}\right)^{1/4} , \tag{1.20a}$$

$$u_\eta \equiv (\nu\varepsilon)^{1/4} , \tag{1.20b}$$

$$\tau_\eta \equiv \left(\frac{\nu}{\varepsilon}\right)^{1/2} . \tag{1.20c}$$

Those definitions imply that the length-scale ($l < \eta$) cannot be sustained since the viscous effects dominates the inertial force (i.e. $Re_l < 1$).

Furthermore, all length-scales significantly smaller than the largest eddy size l_0, then the corresponding turbulent flow motions can be considered to be statistically isotropic (**Kolmogorov's hypothesis of local isotropy**). An illustrative summary of the class of length-scales are shown in in Figure 1.5.

Finally, it is important to highlight that the width of the length-scale spectrum of the turbulent fluid motions is a function of Reynolds number, and can be approximated as a ratio of the smallest to the largest length-scale as:

$$\frac{\eta}{l_0} \approx Re^{-3/4} \,, \tag{1.21}$$

which implies a broader length-scale distribution inside the turbulent flow with higher Reynolds number. Note the consistency between this notion of broader length-scale distribution and increased complexity of the turbulent motions at high Reynolds number that was mentioned earlier in this chapter.

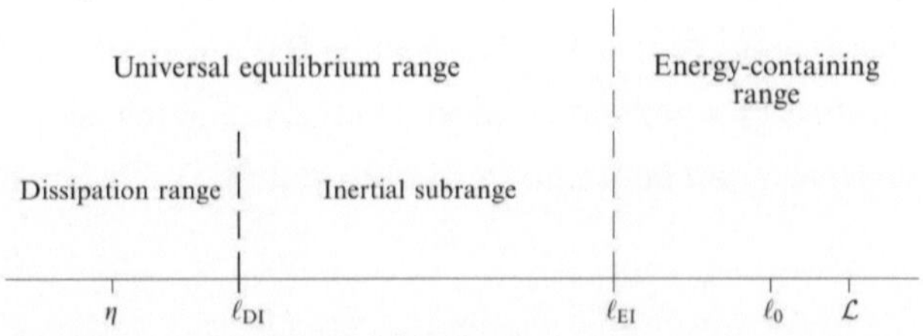

Figure 1.5: Class of the length-scales in the turbulence with very high Reynolds number, plotted in logarithmic scale. Annotations of the axis in ascending order are: η, Kolmogrov length-scale; l_{DI} ($\approx 60\eta$), the upper limit of the dissipation range; l_{EI} ($\approx \frac{1}{6}l_0$), upper limit of the universal equilibrium range and the inertial subrange; l_0, the largest eddy size; $\mathscr{L}$, characteristic length-scale. This figure is from Pope (2000).

1.2.5 Fundamentals of wall-bounded turbulence

At this end of the turbulence introduction before proceeding to the literature surveys, some fundamental aspects specific to the wall-bounded turbulence need to be introduced.
For that purpose, it is convenient to consider a canonical fully-developed *boundary layer flow* that is a stream of fluid flowing over a planar solid-wall surface. It is important to note the generality of the simplest flow configuration, which in fact every kinds of wall flow, including the flow over any objects in a free-stream (e.g. cars, aeroplanes), are some sort of boundary-layer flow.
Furthermore, such simple flow configuration can be characterised by a set of three global parameters, namely boundary-layer thickness δ, free-stream velocity U_0 and kinematic viscosity ν, forming a corresponding Reynolds number as:

$$Re_0 = \frac{U_0 \delta}{\nu} \tag{1.22}$$

Mathematically speaking, the solid-wall that is present in the boundary-layer flow imposes so-called *impermeable* and *no-slip* boundary conditions simultaneously, which in turn creates a thin layer of viscosity-induced retarded flow in the direct vicinity of their surfaces, hence the name of the flow.
The impermeable boundary condition can be expressed as:

$$v|_{y=0} = 0 \,, \tag{1.23}$$

whereas the no-slip boundary condition imposes

$$u|_{y=0} = w|_{y=0} = 0 \,. \tag{1.24}$$

Note that in the following we also employ the subscript "w" to indicate the physical quantities at the solid-wall surface (e.g. $u|_{y=0} = u_w$) for the sake of simplicity.

From the historical perspective, it was Prandtl (1904) who discovered the existence of such layer that is created by the effect of the fluid viscosity. In fact, his boundary-layer theory was the long-awaited knowledge gap to explain the existence of the wall shear stress, providing a conclusive answer to the famous wall-stress paradox existed since the discovery of *inviscid potential flow* theory in 1752 (i.e. **D'Alembert's paradox**). More extensive review on the historical impacts left by the Prandtl's great discovery can be found in Anderson (2005). The mathematical definition of the wall stress τ_w is

$$\tau_\mathrm{w} \equiv \rho\nu \left(\frac{\mathrm{d}\langle u\rangle}{\mathrm{d}y}\right)_\mathrm{w} . \tag{1.25}$$

Once the flow becomes turbulent, there exists, inside the turbulent boundary layer, two-layer structures that are characterised by the difference in the physical quantities that influence the mean profile of the primary velocity component (i.e. $\langle u\rangle$), namely *inner layer* and *outer layer*. As the names suggest, the former is located near the wall, whilst the latter is away from the wall, sharing an in-between layer only if the flow Reynolds number is sufficiently large.
In the inner layer, $\langle u\rangle$ is determined by the velocity- and length-scales that are the solo-products of the viscous wall shear τ_w (i.e. independent of the global quantities, such as δ and U_0), viz.

$$u_\tau \equiv \sqrt{\frac{\tau_\mathrm{w}}{\rho}} , \tag{1.26}$$

and

$$\delta_\nu \equiv \nu\sqrt{\frac{\rho}{\tau_\mathrm{w}}} = \frac{\nu}{u_\tau} . \tag{1.27}$$

Whilst the velocity-scale is often referred to as *friction velocity*, the latter is called *viscous length-scale*. Note that the above distinction of the inner layer from the rest of the flow as well as the aforementioned scaling properties were first postulated by Prandtl (1925), and often referred to as **Prandtl's law of the wall**.
By definition the Reynolds number defined by those scales is at unity (i.e. $u_\tau\delta_\nu/\nu = 1$). Alternatively one can define *friction Reynolds number* Re_τ by the boundary-layer thickness δ, viz.

$$Re_\tau \equiv \frac{u_\tau\delta}{\nu} = \frac{\delta}{\delta_\nu} . \tag{1.28}$$

Correspondingly, let us refer to the velocity and length quantities normalised by those friction scales as *wall units*, and we will denote those quantities by the superscript "+", such as the "wall distance in wall units", viz.

$$y^+ = \frac{y}{\delta_\nu} = \frac{yu_\tau}{\nu} , \tag{1.29}$$

and the "mean (streamwise) velocity in wall units":

$$u^+ = \frac{\langle u\rangle}{u_\tau} . \tag{1.30}$$

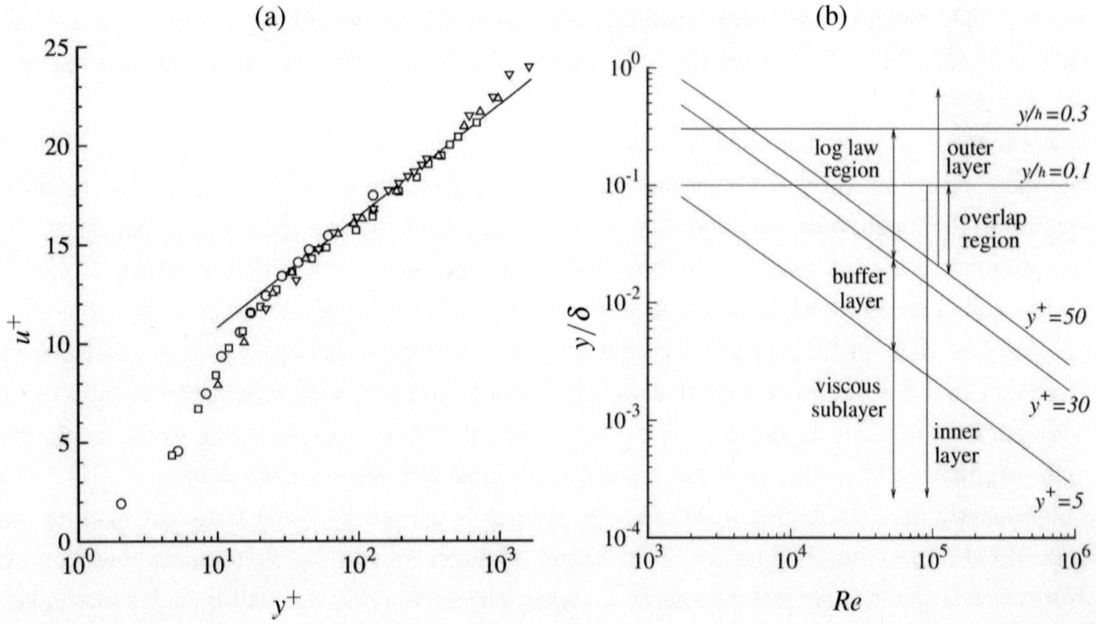

Figure 1.6: (a) Mean velocity profiles in fully developed turbulent channel flow by Wei and Willmarth (1989), with a range of Reynolds numbers $Re_0 = [2970, 14914, 22776, 39582]$. (b) Wall-distance in outer scales of the different regions in turbulent wall-bounded flow as a function of the Reynolds number. Both figures are from Pope (2000).

Based on those notations, the locations of the inner and outer layers can be approximately defined as $y/\delta \lesssim 0.1$ for the former, and $y^+ \gtrsim 50$ for the latter. Note a general consensus that the statistical properties inside the outer layer can be characterised by the global quantities such as U_0 and δ, often referred to as the *outer-scaling* or *bulk-scaling* phenomenon. In contrast, we will refer as the *inner-scaling* or *wall-scaling* phenomenon to those statistical properties observed inside the inner layer which can be characterised by the wall units. Let us now proceed to the detailed organisations inside the inner layer.

Deep inside the inner layer where the normalised wall distance is below 5 (i.e. $0 \leq y^+ \lesssim 5$), there exists a region called *viscous sublayer* that satisfies

$$u^+ = y^+ \,. \tag{1.31}$$

Such linear relation is readily visible in the mean streamwise velocity profile plotted in the semi-logarithmic scale presented in Figure 1.6(a)). Note the presented figure is from the plane-channel flow, but the profile inside the inner layer should not to be affected due to the independence from the outer quantities.

On the other hand, in the region further away from the wall but still inside the inner layer (i.e. $y^+ \gtrsim 50$ and $y/\delta \lesssim 0.1$), there exists so-called the *logarithmic layer* which satisfies

$$u^+ = \frac{1}{\kappa} \log(y^+) + B \,, \tag{1.32}$$

where κ and B are the *von Kàrmàn constant* and an arbitrary constants respectively. This is the **logarithmic law of the wall due to von Kàrmàn**, or simply referred to as the **log law**. The precise values of those constants are the topics of discussion, but widely accepted values are $\kappa = 0.41$

and $B = 5.2$. Note that for the logarithmic layer to be developed fully, one can show from the locational definitions of the layer that it is necessary to have sufficiently high Reynolds numbers (i.e. $Re_\tau > 500$).

In between the viscous sublayer and the logarithmic layer (i.e. $5 \lesssim y^+ \lesssim 50$), there exists an intermediate region where the mean streamwise velocity profile gradually adapts from the linear relation to the logarithmic one, which is generally referred to as the *buffer layer*. An illustrative summary of the layered structures in the near-wall region can be seen in Figure 1.6(b).

To this end of the introduction to the near-wall turbulence, it is important to note, from the physical point of view, that the presence of the no-slip walls means the absence of the translational symmetry in the wall-normal direction. In other words, the geometries and the dynamics of the turbulent fluid motions depend largely on their distance from the neighbouring walls, in addition to the original length-scale dependency that was discussed in the previous section.

Consequently, the distribution of kinetic energy, that is carried by those turbulent motions and transported through their dynamics, is no longer homogeneous in the wall-normal direction. In order to quantitatively discuss the energy transport phenomenon in the wall-bounded turbulence, it is convenient to introduce the *turbulent-kinetic-energy budget* by simplifying Equation 1.13 with the statistical conditions specific to some fully-developed wall flow (including the fully-developed boundary-layer and plane channel). Those conditions are: the statistical stationarity (i.e. $\frac{\partial}{\partial t} = 0$), the spatial homogeneity, i.e. invariance in the streamwise and spanwise directions ($\frac{\partial}{\partial x} = \frac{\partial}{\partial z} = 0$), and zero mean wall-normal/spanwise velocity components. Then the TKE transport equations can be reduced to

$$\mathscr{P} - \tilde{\varepsilon} + \nu \frac{\partial^2 k}{\partial x_j^2} - \frac{1}{2}\frac{\partial \langle v' u'_j u'_j \rangle}{\partial y} - \frac{1}{\rho}\frac{\partial \langle v' p' \rangle}{\partial y} = 0 \,, \tag{1.33}$$

where from the left: production, pseudo-dissipation, viscous diffusion, turbulent transport and pressure transport term respectively.

Moreover, since the wall-normal gradient of the primary streamwise velocity should be present in this flow configuration, the TKE production term Equation 1.12 can be reduced to

$$\mathscr{P} = -\langle u' v' \rangle \frac{\partial \langle u \rangle}{\partial y} \,. \tag{1.34}$$

As a representative example, the term-by-term distribution of the TKE budget in the plane channel DNS of Kim et al. (1987) is presented in Figure 1.7. It is readily visible from the distribution of the TKE budget terms that the buffer layer contributes to the predominant energy production of the overall TKE balance by receiving the energy from the mean velocity field. Geometrically speaking, the flow structures become dominantly one-dimensional (at $y^+ \approx 7$) by the presence of the near-wall vortices and co-existing streaky jet structures that are highly elongated in the streamwise direction.

Since the generated TKE cannot be fully dissipated within the intermediate layer, the excessive energy needs to be transported to elsewhere. It turns out that a large potion of the excessive energy is transported towards the no-slip wall (i.e. the viscous sublayer) by means of the viscous diffusion.

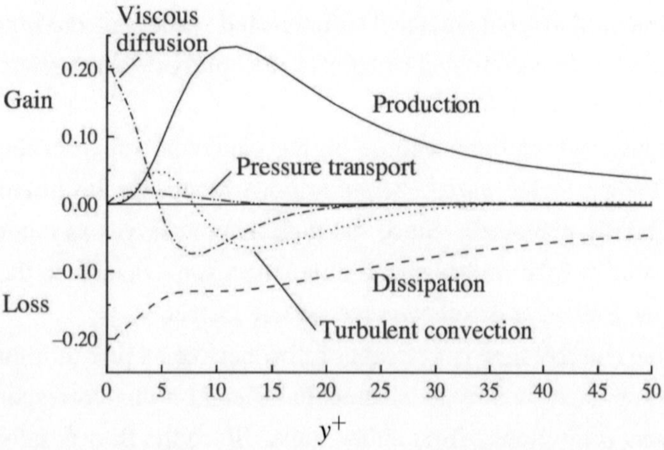

Figure 1.7: The turbulent-kinetic-energy budget in the viscous wall region of plane channel flow with $Re_\tau = 180$ (DNS data from Kim et al. (1987)). All quantities are in the wall units. This figure is from Pope (2000).

Inside the viscous sublayer, the turbulent flow structures are in general very small, and flat in shape due to the presence of the impermeable wall in the direct vicinity which strongly restricts any wall-normal fluid motions. As a result, the fluid viscosity plays a predominant role by dissipating the kinetic energy of those small-scale motions into heat, being by-far the most significant energy sink location globally.
The remaining part of the excessive energy is transported away from the wall towards the outer layer, where the energy dissipation is superior to the energy production which reaches eventually to zero (not shown in Figure 1.7). In between, the logarithmic layer exists where the energy production and dissipation are approximately in balance.
The general consensus upon the geometric interpretation of the logarithmic layer turbulence is still to be established, in contrast to the counterparts existing in the layers considerably nearer to the walls, and it will be mentioned during the following literature survey.

1.3 Literature survey

1.3.1 Wall-bounded turbulence

We have finally reached to this point where we can explore the state-of-the-art of the topics of concern, and we will start with the fundamental aspects of the wall-bounded turbulence first.
Since the topic has been investigated extensively over the last decades, there exists a vast number of references. We are therefore required to limit our focus almost exclusively on the study of the turbulent coherent structures reside below the buffer-layer limit of the canonical wall-bounded flow. The readers with broader interests than the scope of this survey are encouraged to refer to the more extensive reviews on the subject (e.g. Jiménez (2013); Jiménez and Kawahara (2013)).
Furthermore, it is important to note that the considered wall-bounded flow is with marginal to moderate Reynolds number, which imply that those flow may or may not have developed the logarithmic layer with its full extent. Such Reynolds number restriction arises more from the availability of the data with sufficiently high temporal and spatial resolutions, rather than from

the design of the structural investigations. The interested readers of the high Reynolds number turbulence are therefore also encouraged to refer to the appropriate materials, such as the recent review of the topic by Smits et al. (2011).
For the current purpose, we will concentrate on the canonical fully-developed turbulent plane channel flow, also known as the *plane Poiseuille flow*, alongside the turbulent boundary-layer flow that had been briefly explained. Since the type of flow serves as one of the most general configuration in the study of the wall-bounded turbulence, some details on the configuration need to be introduced in the following before we proceed any further.
Fully-developed plane channel flow is a stream of flow between a pair of infinitely long and wide no-slip planes, often characterised by its channel half-height h that corresponds to the thickness of the boundary-layers δ developed from either walls. Since the flow is fully-developed and the spanwise extent is assumed to be infinite, the flow statistics become homogeneous in both the streamwise and spanwise directions; in other words the statistics are one-dimensional therefore only functions of the wall distance y. In order to realise such flow configuration, periodic boundary condition is employed in the two directions in a case of numerical simulations. Conversely in the experimental investigations, flow through a rectangular duct with large aspect ratio ($A \gtrsim 25$, where A is the duct aspect ratio. cf. Nishioka and Asai (1985); Vinuesa et al. (2015)) is employed, and the flow statistics are accumulated only over the duct core section where any side-wall effects are considered to be negligible.
Whilst the definitions of the characteristic friction-based quantities that explain the statistics and dynamics of the near-wall viscosity-dominant regions remain identical, the boundary-layer free-stream velocity U_0 that characterises the largest scales of fluid motions becomes the centre-line velocity instead in a case of plane channel flow. Furthermore, one can define an additional velocity scale that explains the outer-scaling nature of the flow statistics, namely *bulk velocity* as:

$$u_b = \frac{1}{2h} \int_{-h}^{+h} \langle u \rangle dy \,, \tag{1.35}$$

and the corresponding *bulk Reynolds number* can be defined as:

$$Re_b = \frac{u_b h}{\nu} \,. \tag{1.36}$$

The study of the near-wall turbulent coherent structure generally requires extremely high temporal and/or spatial resolutions of multidimensional flow data due to the small-scale (i.e. inner-scaling) natures of those structures. Such severe requirements had been a major limiting factor for our understanding until the end of the second-last decade of the previous century, since the available data were limited to the experimental pointwise measurements (e.g. hot-wire anemometry, hot-film anemometry, Laser-Doppler anemometry), and the flow visualisation techniques based on the smoke injected into the wind tunnels and the hydrogen bubbles in the water tunnels. Nevertheless, those techniques successfully revealed some of the fundamental structures and their dynamics observed in the near-wall regions, namely *velocity streaks*, *ejections* and *sweeps* (cf. Kline et al. (1967); Kim et al. (1971)); and built a theoretical foundation necessary for the forthcoming era of direct numerical simulations (e.g. plane channel DNS by Kim et al. (1987); boundary-layer DNS by Spalart (1988)).

The velocity streaks are the streaky structures predominantly observed inside the viscous sublayer and the following buffer layer, with their geometry being highly elongated in the streamwise direction whilst compact in both other directions. Those streaks can be seen as arrays of the high and low-velocity jets readily visible as the *positive* and *negative* u' regions, alternating their positions in the spanwise direction. Note the former is referred to as *high-velocity streaks*, whilst the latter is referred as *low-velocity streaks*. The streamwise extent of the velocity streaks is in $\mathcal{O}(1000\delta_\nu)$, whereas the average spanwise spacing between the streaks of opposite sign is approximately $50\delta_\nu$, which implies the average spacing between the same-signed streaks (e.g. distance from a high-velocity streak to the closest high-velocity peer) to be around $100\delta_\nu$ (cf. Smith and Metzler (1983); Kim et al. (1987)).

The velocity streaks are often associated with the *quasi-streamwise vortices*, tilting on average 4 degrees in xy-plane and 9 degrees in xz-plane (note: therefore not strictly *streamwise*, but *quasi-streamwise*), spanning approximately $200\delta_\nu$ in the streamwise direction with the diameter of $\approx 25\delta_\nu$ (cf. Jeong et al. (1997)). The average streamwise spacing of those vortices was found to be about $300\delta_\nu$ (cf. Jiménez and Moin (1991)).

The appearance of the streaks can be triggered by imposing high mean shear rate, therefore independently from the explicit existence of the no-slip walls (cf. the homogeneous shear flow DNS of Lee et al. (1990)). Those streaks are the consequence of the pairs of counter-rotating quasi-streamwise vortices advecting the near-wall low-momentum fluid upwards (i.e. positive v'), and the continuity constraint enforce the opposite downward motion of the high-momentum fluid towards the no-slip walls (negative v') to be formed in the adjacent locations (cf. Blackwelder and Eckelmann (1979)).

Both phenomenon are associated with the negative values of the off-diagonal Reynolds stress component $u'v'$ and contribute predominantly to the energy production term (cf. Equation 1.34). If we project those motions onto the u'-v' sample space, the upward motions of the low-velocity streaks are categorised in the second quadrant of the space, whereas the downward motion of the high-velocity streaks are categorised in the fourth quadrant. Consequently, the former is generally referred to as *"ejection"* (or alternatively *Q2*) events, whereas the latter is called *"sweeping"* (*Q4*) events, or simply *"sweeps"*.

The related term *"bursting"* was initially employed to describe the intermittent break-up of the low-velocity streaks observed experimentally by, for example, Kline et al. (1967). The existence of such intermittent phenomenon was then questioned even by those original authors (cf. Offen and Kline (1975)), and it was once concluded that their earlier observations could be an artefact of the pointwise measurements, misinterpreting the advected sweeps and ejections occasionally passing through the measurement probe as the conjectured intermittent events (cf. Robinson (1991)).

The direct numerical simulations of the minimal plane channel (i.e. with the numerical domain size just large enough to contain the minimal unit of the turbulent coherent structures to sustain the turbulence) by Jiménez and Moin (1991), however, overturned the almost-formed consensus, by demonstrating that there exists indeed the quasi-periodic bursting phenomenon separated by the typical time-interval of $\tau_B^+ = \tau_B u_\tau^2/\nu \approx 400$. It was later confirmed that such bursts can be also identified in the full-size direct numerical simulations, if we consider the flow dynamics inside the randomly-chosen the minimal-unit sub-domains (cf. Jiménez et al. (2005)).

In relation to the bursting phenomenon, it was proposed by Swearingen and Blackwelder (1987) that there exists a regeneration cycle involving the low-velocity streaks and the quasi-streamwise vortices through an inflectional instability of the streaks. Subsequently, Jiménez and Pinelli (1999) demonstrated by their numerical experiments with the outer flow fluctuations artificially filtered, that there exists a dominant autonomous regeneration cycle whose sustainability is independent from the rest of the flow where $y^+ \gtrsim 60$. Within the regeneration cycle, the streaks and vortices are created, grow, generated each other, and eventually decay; and the inflectional instability that is triggered by sinuous disturbances play a crucial role (cf. Jiménez (2013), and their Section IV for other references).

The propagation velocities of those near-wall coherent structures are also interesting aspects to be mentioned. Kim and Hussain (1993) analysed the *x-t* correlation of the plane channel flow with $Re_\tau = 180$ computed by Kim et al. (1987), and concluded that the advection velocities are approximately equal to the local mean velocities, except the direct vicinity of the no-slip walls $y^+ \lesssim 15$ where the advection velocity was found to be approximately constant which is significantly higher than the local mean. Kim and Hussain (1993) explained their observations by a suggestion that those structures are predominantly the signatures of the larger structures that exist farer away from the walls. The appropriate velocity scaling could not be deduced then since they only considered one Reynolds number flow. del Álamo and Jiménez (2009) later computed the velocities spectrally for the plane channel flow with various Reynolds number $180 \leq Re_\tau \leq 1900$, and supported the earlier findings with the constant advection velocities of below $y^+ \approx 15$ being $\approx 11u_\tau$. Moreover, del Álamo and Jiménez (2009) found that the large-scale structures, that are proportional to the global dimensions of the flow domain, travel also at a constant velocity approximately equals to the bulk flow velocity.

To this end, it is important to discuss some of the dominant flow features observed in the logarithmic layer, at least briefly to conclude this survey of the near-wall turbulence. As it was mentioned earlier, though, our understanding in the topic is still under rapid development and extensive investigations are currently conduced throughout the world. It is therefore reasonable for us to cover only the aspects that are well developed by now, and will be relevant to our duct flow investigations with the marginal Reynolds number.

One of the two most generally-accepted coherent-structure models to describe the logarithmic-layer flow is the family of *hairpin vortex* including: the *horseshoe*, Λ, Ω and *cane* vortices that have been proposed by different authors over the last years with slightly different details, but can be categorised into one (cf. Adrian et al. (2000)). The model was initially proposed by Theodorsen (1952), and later caught much of attention by being demonstrated by the smoke visualisation of Head and Bandyopadhyay (1981).

The simplest hairpin vortex model consists of two parts: two counter-rotating vortex *legs* that are extended upwards from the neighbouring wall, and the *head* or *arch* connecting those legs and often located in the logarithmic layer and beyond. Note that those legs are some of the buffer-layer quasi-streamwise vortices but not all of them form the hairpin vortices. Note also that many hairpin vortices are not even approximately symmetric in shape and some of them only have one leg, which are occasionally referred to as *vortex canes*.

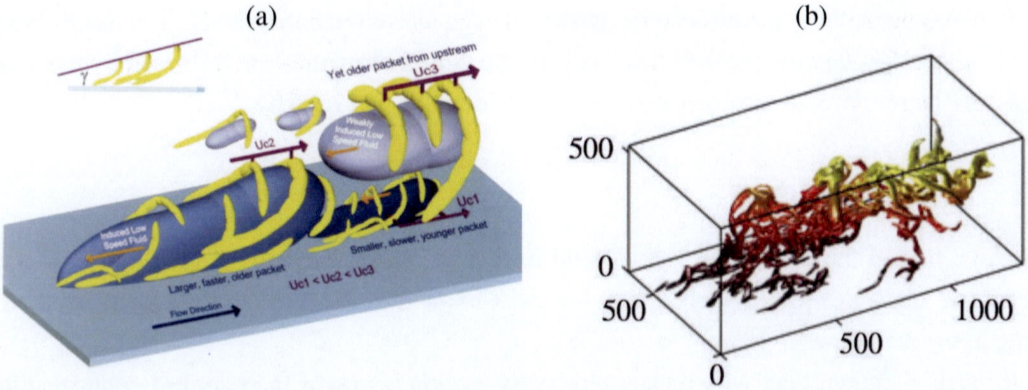

Figure 1.8: (a) Conceptual scenario of nested hairpin vortex packets. This figure is from Adrian et al. (2000). (b) An example of wall-attached self-similar vortex cluster, coloured by the wall-normal height. This figure is from Lozano-Durán et al. (2012). In both figures, the streamwise flow direction is from left to right.

It is possible to observe the isolated hairpin vortices in the near-wall turbulence, though more frequently they form *hairpin vortex packets*, through the auto-generation mechanism triggered by certain conditions (cf. Zhou et al. (1996); Zhou et al. (1999)). Those hairpin packets are often aligned in the streamwise direction, one after the other. The individual hairpin packets grow steadily in the wall-normal direction as well as the spanwise direction as they age, aligned in space in the former direction at a mean angle of approximately 12 degrees (cf. Adrian et al. (2000), also Figure 1.8(a)). Those older and taller packets are advected faster than the younger and shorter packets, occasionally run over the younger peers existing closer to the walls, and eventually forming a hierarchical organisation of the self-similar hairpin vortices.

The other type of the coherent-structure model, though it is not necessarily irrelevant to the above hairpin vortex paradigm, is the *wall-attached eddy* model proposed by Townsend (1961), and later detailed further by Perry and Chong (1982). Note that we distinguish between *vortex* and *eddy* here to indicate that the former is in the order of Kolmogorov length-scales, whereas the latter object is large-scale and energy-containing (cf. Jiménez and Kawahara (2013)). The core idea is that there exists two types of the large-scale vortical motions: tall structures attached to the no-slip walls, highly anisotropic in shape and carrying a large portion of Reynolds shear stress; and the other structures detached from the walls (i.e. *wall-detached eddy*) and more isotropic in shape and carrying a little portion of the shear stress. Note again that this concept can be consistent with the hairpin vortex model by considering that those packets can be attached and detached from the no-slip walls.

del Álamo et al. (2006) proposed a less organised, but perhaps therefore more general model to refine the classical model of Townsend (1961), namely *self-similar vortex clusters*. They are the shells of disorganised vortices extending beyond the buffer-layer (cf. Figure 1.8(b)), with their width and length proportional to the height (i.e. scaling with the wall distance), with the estimated aspect ratios of $l_x \times l_y \times l_z = 3 \times 1 \times 1.5$.

The main difference between the hairpin vortex packet and the self-similar vortex cluster models is, apart from the degree of the geometrical coherence, that the former type of the structures have to be generated in the near-wall region through the aforementioned regeneration cycle and grow from there, whereas the latter structures can be generated even above the buffer-layer. The difference is rooted from the fact that the regeneration cycle of the cluster structures are triggered by the *wake* behind their larger peers in the downstream (cf. del Álamo et al. (2006)).

1.3.2 Duct turbulence

In this final part of the current survey, we will review the state-of-the-art of the duct mean secondary flow. Once again, our special attention will be paid to the turbulent coherent structures existing in the duct flow.

Flow through a straight duct with rectangular cross-section is one of the simplest geometry that experience turbulence-driven mean cross-stream motions, commonly known as "*secondary motion of Prandtl's second kind*". Existence of such secondary motions has been known for long since the early measurements by Nikuradse (1926), and more recent experimental and numerical studies revealed their precise geometrical features: a pair of counter-rotating mean streamwise vortical motions full-filling each corner, transporting the high-momentum fluid from the duct core into the corner regions along the cross-stream plane diagonals (cf. Figure 1.2(a)). Despite their weak intensity of few percent of the bulk velocity, the secondary motions are known to play a significant role in the momentum, mass and heat transfers through a modification of the primary streamwise velocity distribution.

Significant amount of effort has been devoted to explore the origins of the duct secondary motions, both experimentally (e.g. Baines and Brundrett (1964); Gessner (1973); Melling and Whitelaw (1976)), and later numerically by means of direct numerical simulations (cf. Gavrilakis (1992); Huser and Biringen (1992)). Many of those early studies focused on the mean streamwise vorticity budget derived from RANS equations (cf. Equation 1.18), and demonstrated a sustaining mechanism of the mean vorticity field by the mean Reynolds stresses balance. Those studies highlight the main cause of many of the common turbulence models' failure to predict the secondary motions even qualitatively (cf. Raiesi et al. (2011)), and promote the direct numerical simulation techniques, which do not depend on any turbulence model, as a favourable research tool.

More recently, a number of numerical studies shed a spotlight on the dynamics of the near-wall turbulent coherent structures to exploit the full spatio-temporal resolutions available from the direct numerical simulations.

Uhlmann et al. (2007) studied square duct flow for a range of marginal Reynolds number in the limit of minimal streamwise extension, and determined the minimal sustainable friction-based Reynolds number of the flow at $Re_\tau \approx 80$ ($Re_b \approx 1100$). Note that the friction Reynolds number $Re_\tau = u_\tau h/\nu$ is based on the the friction velocity based on the wall stress average over all no-slip walls, and the duct half-height h. Geometrically speaking, the discovered critical Reynolds number refers to the full duct height normalised by the friction length scale $(2h)^+$ just long enough to host three buffer-layer velocity streaks that consist of a low-velocity streak located at each wall's bisector flanked by a pair of high-velocity streaks. Their extensive vortex-eduction studies completed the new picture by revealing the high probability of existence of a pair of counter-rotating streamwise vortices sustaining the in-between low-velocity streak on each edge of the square duct, coinciding very well with the well-known eight-vortex profile.

Furthermore, they also reported the existence of temporally-alternating four-vortex mode embedded within the duct flow at the marginal Reynolds number of lower than $Re_b \approx 1250$, satisfying the eight-vortex profile only after averaging over a substantial interval in space (along the streamwise homogeneous direction), and in time. Instantaneously, only one adjacent wall pair hosts a low-velocity streak on each edge, while turbulent activities on the other pair of walls remain significantly quiescent. Note that similar alternating phenomenon in minimal wall-bounded turbulence was reported previously by Jiménez and Moin (1991) for the plane channel flow.
Subsequently, their findings of the direct relation between the preferential locations of the turbulent coherent structures and the long-term statistics were further strengthened by a family of invariant travelling-wave solutions found by Uhlmann et al. (2010). Their travelling-wave solutions with simple staggered streamwise vortices and wavy streaks arrangement capture the famous eight-vortex profile as well as the integral quantities of the long-term averaged fully-developed turbulent statistics surprisingly well, implying their dynamical importance in the duct turbulence.
Pinelli et al. (2010) performed a series of direct numerical simulations up to $Re_b = 3500$ ($Re_\tau \approx 230$), and found that their nature of hosting an integer number of velocity streaks and the corresponding preferential locations of the turbulent coherent structures imposed by the duct geometry still determine the mean statistics for this range of Reynolds number.
Furthermore, it was also found that the probabilistic distributions of low- and high-velocity streaks become progressively channel-like uniform along the duct walls by recovering the translational symmetry with increased Reynolds number, except in the direct vicinity of each corner. In the corner region, a high- and low-velocity streak pair maintains constant distances in wall units from the closest corner (lesser extent for the low-velocity streak than the high-velocity one) independent of Reynolds number.
Pinelli et al. (2010) also studied the scaling properties of the mean secondary velocity and the streamwise vorticity patterns with a special emphasis upon their peak locations with respect to the nearest walls. They found that the peak locations of the eight-vortex secondary velocity profile depend solely on the bulk geometry, while the streamwise vorticity peaks move towards the corners with increased Reynolds number, maintaining constant distances from the corner in wall units (as the velocity streaks pair). Consequently, the mean secondary velocity and the mean streamwise vorticity profiles differ progressively from each other for higher Reynolds number, witnessing some degree of scale-separation and the significant contribution from the emerged large-scale motion to the mean secondary flow pattern at this moderate Reynolds number range.
More recently, Takeishi et al. (2015) conducted a systematic investigation of the transitional closed duct turbulence with a range of aspect ratios $A = \{1 \ldots 9\}$ (note: here A is defined as the ratio of the duct half-span to the duct half-height h), by means of the pseudo-spectral direct numerical simulation technique employed by Uhlmann et al. (2007) and Pinelli et al. (2010). Note that to capture the spatially localised feature of the transitional turbulence, the streamwise extent of their numerical domain ($L_x = 40\pi/h$) was set to be significantly longer than the one employed by Uhlmann et al. (2007) and Pinelli et al. (2010) ($L_x = 4\pi/h$).
They reported an interesting aspect ratio dependence in the minimal sustainable Reynolds number, approaching continuously to the lowest limit observed in the plane channel flow, as the duct aspect ratio increases.

Takeishi et al. (2015) also observed a fundamental structural change in the marginal duct turbulence, with respect to a critical aspect ratio of ≈ 4. They observed that the transitional duct flow exhibits streamwise localised puff similar to the one observed in the turbulent pipe flow, when the aspect ratio is smaller than the critical value. On the other hand, the turbulent structures detached from the duct side-walls and eventually form oblique bands of the turbulent structures, more similar to the flow features observed in the transitional plane channel flow. Takeishi et al. (2015) demonstrated that the structural detachment is a consequence of the localised relaminalisation in the side-wall regions, caused by the wall stress reduction observed in the regions within the wide aspect ratio ducts.

A variant of the rectangular duct, with one of the four enclosing solid walls being replaced by a free surface, has an additional range of prominent applications such as man-made canals and urban rivers. As it was mentioned earlier, we refer to this kind of configuration as "open duct", to distinction from the configuration without the free surface (i.e. "closed duct"). Note that the characteristic length-scale of the flow is the duct full-height H, in contrast to the closed duct case whose equivalent length-scale is h.

Turbulent flow through an open duct is known for long to exhibit a secondary flow pattern that is very different from the closed duct counterpart (cf. Nezu and Rodi (1985); Tominaga et al. (1989)), partially characterised by a pair of large mean streamwise vortices extended from each side of the walls towards the duct centre.

Such mean free surface vortices were found to extend up to twice of the full duct height (i.e. $2H$) in the spanwise direction in a case of high aspect ratio duct flow, whereas the mean bottom-wall vortices, which are similar to the ones observed in the closed duct flow, extend only up to H in the same direction. Those experimental observations imply that the mean secondary flow structures remain relatively close to the duct side-walls, even with an aspect ratio which is significantly larger than unity. Note that the side-wall localised nature of the mean secondary motions in the wide rectangular closed ducts were recently confirmed numerically, by the direct numerical simulations of Vinuesa et al. (2015).

In case of flow through narrow ducts, the mean free surface vortices are considered to be responsible for shifting the location of the maximum streamwise velocity somewhere beneath the free surface even at the duct mid-span. Such phenomenon is often referred to as *"dip-phenomenon"*, and it has been known since even more than one century ago (cf. Stearns (1883)). More recently, the experimental investigations of Tominaga et al. (1989) reported that the submerged distance is about $0.2-0.3H$ from the free surface at the duct mid-span, if the duct aspect ratio is less than or equals to a critical value (i.e. $A \lesssim A_{\text{crit}} \approx 2.5$. Here A is the open duct aspect ratio, defined as the ratio of the duct half-width to the duct full-height H).

Although the phenomenon has a high engineering relevance with a long history of investigations (e.g. Tracy and Lester (1961); Sarma et al. (1983); Sarma et al. (2000); Wang et al. (2001); Yang et al. (2004)), its precise physical mechanism is still lacking from our knowledge. Also to our knowledge, possible consequence of the Reynolds number dependence of the phenomenon has never been addressed previously.

In spite of the high technical relevance, however, our physical understanding of the open duct flow is much more limited in comparison to the ones achieved for the closed duct counterpart to date.

Technical difficulties to conduct accurate measurements in open duct experiments —mainly due to the fact that water is often used as working fluid in contrast to the closed duct experiments which can be done with air, in addition to the challenges involving free surface treatment— have been the main issue to cause such slow progress. For historical aspects of the experimental challenges and their attempted solutions, the interested readers are encouraged to consult Nezu and Nakagawa (1993) and Nezu (2005).

Consequently, a complete picture of the secondary flow pattern became available only recently. Extensive experimental and numerical studies by Grega et al. (1995) were the first of its kind to reveal the existence of a relatively small mean swirl with opposite sign of rotation with respect to the neighbouring free surface vortex, residing near the mixed-boundary corners formed by free surface and a solid side-wall (cf. Figure 1.2(b)). The mixed-boundary corner vortex was named "*inner secondary flow*" while the free surface vortex was referred to as "*outer secondary flow*" by the authors. They also provided new insights into the dynamics of the turbulent coherent structures near the mixed-boundary corners, such as bursting-like phenomenon directly below the free surface originating from the duct side-walls; occasional appearance of filament-like structures again from the side-walls towards the duct core; and a low-velocity streak consistently located on the side-walls approximately 50 wall units below the free surface.

Follow-up experimental investigations were conducted with improved resolutions in order to study the averaged transport phenomena of turbulent kinetic energy (cf. Hsu et al. (2000)); and the streamwise vorticity (cf. Grega et al. (2002)) in the mixed-boundary corner regions.

A number of relevant numerical studies had been also performed and they reproduced the newly-discovered picture by means of large-eddy simulations (cf. Thomas and Williams (1995); Sreedhar and Stern (1998); Broglia et al. (2003)) and more recently by direct numerical simulation (cf. Joung and Choi (2010)), though in a relatively short streamwise domain length with coarse resolutions. Note that the free surface deformation is neglected in the aforementioned numerical studies which imply zero Froude and zero Weber number limit of the flow. Note also that, to our knowledge, fully-resolved open duct direct numerical simulation with a sufficiently long domain has not been performed previously. Furthermore, an extensive study on the turbulent coherent structures up to the level where it has been already applied to the closed duct flow is still to be done for the current open duct configuration.

1.4 Objectives and organisation of this thesis

Through the above literature survey, the following key knowledge gaps in the study of open duct turbulence were identified:

1) key coherent structures and their dynamics responsible for the open duct secondary motions;

2) existence of any clear marginal flow features, such as the temporally-alternating four-vortex state observed in the closed duct flow;

3) Reynolds number and aspect ratio dependence in the mean secondary flow patterns in the fully-developed turbulence regime;

4) Reynolds number and aspect ratio dependence in the dip-phenomenon;

5) both qualitative and quantitative similarity and difference with respect to the closed duct secondary motions.

In order to answer the above questions, *high-fidelity* dataset should be generated by fully-resolved open duct direct numerical simulations performed in long-enough simulation domains. The simulated Reynolds number range should cover from the marginal (i.e. $Re_\tau \approx \mathcal{O}(100)$) to the moderate (i.e. $Re_\tau \approx \mathcal{O}(400)$) limits, whereas the upper limit of the aspect ratio should exceed significantly the dip-phenomenon critical aspect ratio of ≈ 2.5. We will consider the aspect ratios up to 8 in this thesis. The corresponding closed duct simulations should also be performed for comparison. The rest of this thesis are organised as follows. First, the pseudo-spectral direct numerical simulation technique employed throughout this thesis will be detailed in the following Chapter 2. Subsequently, the Reynolds number dependence in the open duct flow with a fixed aspect ratio $A = 1$ will be investigated in Chapter 3, whereas the aspect ratio dependence in both the marginal and moderate Reynolds number open duct flow will be discussed in Chapter 4. Through those investigations, the key coherent structures responsible for the open duct secondary flow will be identified, and their scaling properties will be discussed accordingly. In Chapter 5, the dynamics of the identified key coherent structures will be investigated by means of a temporal-tracking technique developed for this thesis. Note that such dynamical study of the coherent structures will be the first of its kind applied to the duct flow in general. Finally, the overall summary and outlook of the current work will be given in Chapter 6 to conclude this thesis work.

Chapter 2

Numerical procedure

2.1 Introduction

In this chapter, we shall introduce the pseudo-spectral direct numerical simulation technique employed for this thesis. The simulation code had been developed prior to the current project, and employed in the several closed duct flow studies previously (i.e. Uhlmann et al. (2007), Pinelli et al. (2010), Sekimoto et al. (2011), Takeishi et al. (2015)). It was then extended to incorporate the free-slip boundary condition for the purpose of the current investigation, again prior to the beginning of this project.

The code had been validated against the closed square duct numerical results of Gavrilakis (1992) as well as the experimental measurements of Kawahara et al. (2000) (in Japanese), and the results were detailed in Pinelli et al. (2010) alongside some implementation details. A series of the grid-convergence tests was conducted for the reference open duct case with $A = 1$ and $Re_b = 2205$, and the details will be discussed in Chapter 3 (cf. also Appendix A). The adequateness of the streamwise domain extent will be also examined by means of the standard two-point spatial correlation analysis.

In the following, the overall framework will be summarised in the next section first, which will be followed by the details of the spatio-temporal discretization schemes, and subsequently the employed Helmholtz solvers will be detailed. At this end, it is important to note that the last two parts of this chapter are largely based on the technical report written by Uhlmann and Pinelli (*unpublished*).

2.2 Code overview

We consider incompressible viscous flow through an infinitely long straight open duct with a rectangular cross-section, defined within the Cartesian coordinate system with (x, y, z) being the streamwise, free-slip plane normal, and spanwise directions respectively.

Our numerical domain is defined by the domain lengths in x, y, z directions, each denoted by L_x, L_y, L_z respectively. The streamwise open duct domain length is fixed at $8\pi H$ throughout this study, whilst the cross-sectional dimensions are defined as $[0, H] \times [-AH, +AH]$, where H is the

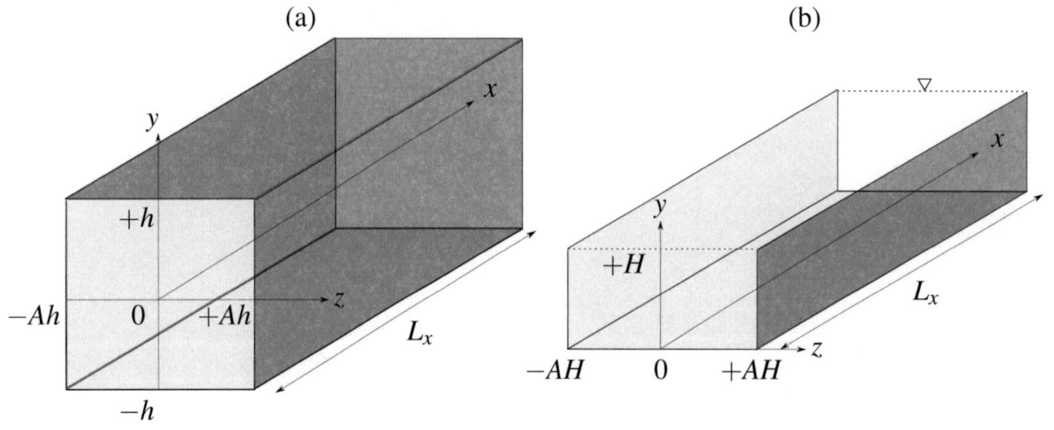

Figure 2.1: Coordinate system and geometry of: (a) closed; and (b) open duct. Shaded planar boundaries represent no-slip walls, whereas the transparent plane enclosed by dotted lines in (b) represents free-slip boundary.

duct full-height, and A is the duct aspect ratio defined as the ratio of the duct half-width to H (cf. Figure 2.1).

For convenience, the boundaries at $z = -AH$ and $+AH$ will be referred to as the (duct) side walls, whilst $y = 0$ will be called the (duct) bottom wall. No-slip boundary condition is imposed on those three boundaries. On the contrary, the boundary at $y = H$ will be referred to as the duct free-slip plane, implying to the fact that free-slip impermeable boundary condition is applied on the boundary. The movement of the free-slip plane is neglected, which corresponds to the zero Froude and zero Weber number limit. Finally, the periodic boundary condition is imposed in the streamwise direction, and a prescribed constant mass-flow rate is maintained during a course of simulations. The definition of A was designed in such a way that the hydraulic diameter ($D_\mathrm{H} = 4\mathscr{A}/P$, where $\mathscr{A} = 2AH^2$ is the duct cross-sectional area and $P = 2H(1+A)$ is the wetted perimeter) is identical to the one for the closed ducts with the same aspect ratio.

We solve the incompressible Navier-Stokes equations (cf. Equation 1.2) for the three velocity components u, v, w defined in x, y, z directions respectively, as well as pressure p normalised by a unit density. Those variables are expanded in terms of truncated Fourier series in the streamwise direction on equidistant grid points, whilst Chebyshev polynomials are used in the two duct cross-sectional directions on collocated *Chebyshev–Gauss–Lobatto* (*CGL*) points. The fields are transformed back and forth by means of *Fast Fourier Transform* (or *FFT*) and *Fast Cosine Transform* (*FCT*). The non-linear convective terms are evaluated in the physical space, while the linear viscous diffusive and pressure gradient terms are evaluated in the spectral space, and aliasing error is removed according to the conventional $2/3$-rule in the Fourier direction only.

A fractional step method is employed in order to decouple the momentum equations from the continuity constraint. The temporal integration is based on the implicit Crank–Nicolson scheme for the viscous terms and an explicit three-step low-storage Runge–Kutta method for the non-linear terms (cf. Section 2.3).

The 2D Helmholtz and Poisson problems for each Fourier mode are solved by a fast diagonalisation technique, and the details are available in Sekimoto (2011). Nevertheless, those details will be repeated in Section 2.4 for completeness.

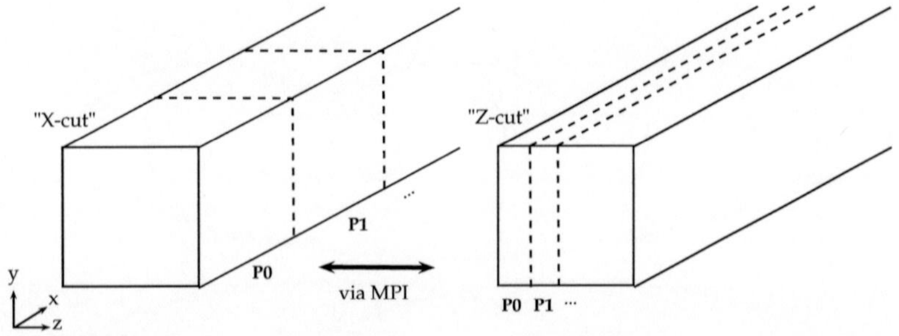

Figure 2.2: "Slice" data decomposition strategy used for parallelisation of the pseudo-spectral DNS code. This figure is from Sakai and Uhlmann (2016).

All simulations presented within this thesis satisfies the following resolution criteria: CFL number is below 0.3; Δx^+ is below 15.2; the number of Chebyshev mode ensures that the maximum cross-stream grid spacing $\max\{\Delta y^+\}$ and $\max\{\Delta z^+\}$ are below 4.1. During the course of the higher Reynolds number simulations of the closed duct flows ($Re_b > 3500$), it was found that the spatial resolution requirements become more severe for the simulations to be performed stably (the corresponding CFL constraint remained approximately unchanged). At the time of writing, the numerical instability is suspected to be due to the spurious pressure modes, arisen from the collocated grid arrangement in the two-dimensional Chebyshev expansions and exist in each Fourier wavenumber, interacting with the no-slip walls and triggering the amplification of the erroneous slip-velocity on those boundaries.

In order to ease the severe requirements, the improved fractional step method proposed by Hugues and Randriamampianina (1998) was implemented during the course of this thesis work, together with the necessary Helmholtz solver for the inhomogeneous Neumann boundary problems (cf. APPENDIX C and D). Although the implementation was validated successfully against the original test cases provided in the reference, and specifically reproduced the promised third-order temporal accuracy in velocity on the boundaries, the method was found ineffective to solve the above numerical instability issue therefore had not been used for any production runs. Note that an alternative two-dimensional vorticity-streamfunction formulation, which does not contain any projection operations therefore free from the slip-velocity error, was also implemented during the course of this thesis work as an experiment, and the details including the validation results are provided in APPENDIX E.

The current DNS code is MPI parallel and the parallelisation of the pseudo-spectral algorithm is achieved by a global data transpose strategy. A schematic of the data decomposition by slices is shown in Figure 2.2 and a cyclic communication pattern, which avoids communication cascade and used for global data transpose, is illustrated in Figure 2.3.

The cost of such communication is $2 \times$ (number of parallel process $- 1$) MPI send/receive calls per data transpose per MPI process. Asynchronous communications are used to minimise the communication overhead. It should be noted that this parallelisation strategy is standard in spectral methods applied to plane channel flow, and more implementation details are available in the PhD thesis of Sekimoto (2011).

In order to demonstrate the efficiency of our DNS code, we have carried out strong scaling tests on CRAY XC40 HORNET at HLRS (cf. Table 2.1, Figure 2.4). The results show that for the chosen

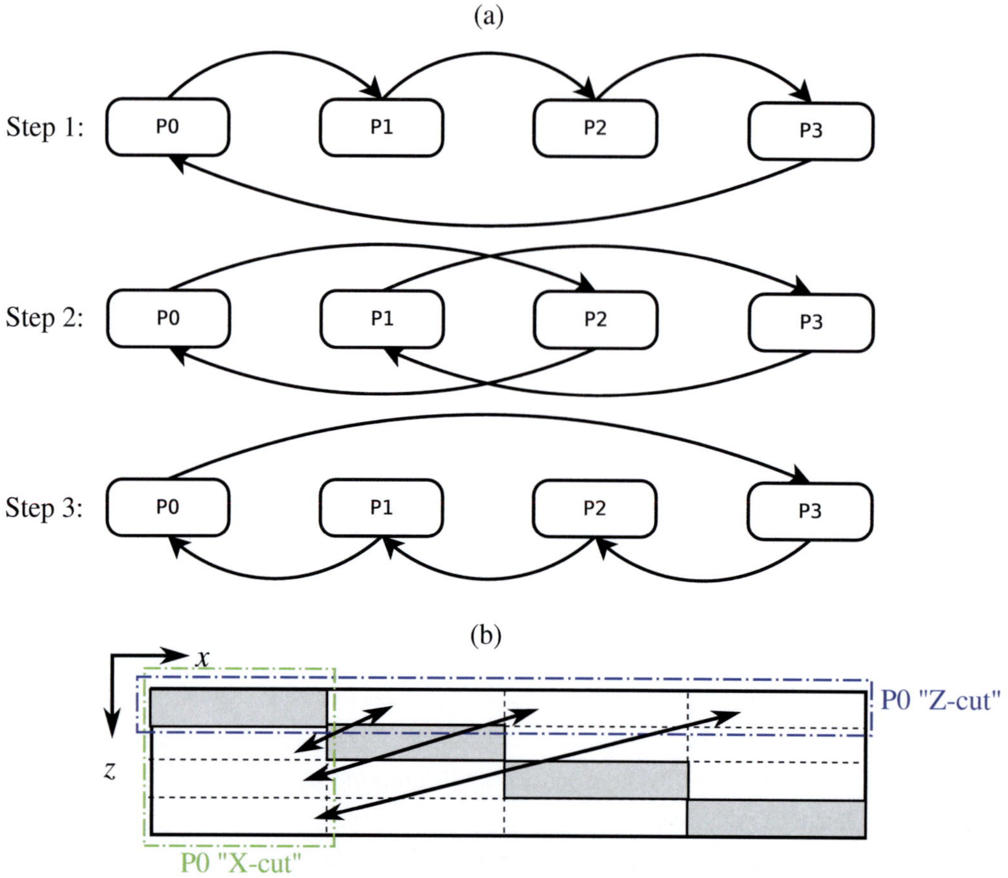

Figure 2.3: (a) A cyclic all-to-all communication pattern used for data transpose between "X-cut" and "Z-cut" data decompositions (cf. Figure 2.2), illustrated by an example of parallel execution model with 4 MPI processes. Each "P*" element represents individual MPI rank, while arrows show data flow by MPI communications. (b) Top view of the above example's domain. Sub-domains surrounded by solid lines (coloured in light gray) are the overlapped regions between "X-cut" and "Z-cut" of each MPI process, hence no communication is required. Any other areas are required to be communicated to other MPI processes. An example of such communications for the MPI process "P0" is illustrated. This figure is from Sakai and Uhlmann (2016).

problem size (approx. 10^8 modes), the code maintains a good parallel efficiency up to several hundred cores. With roughly 10^6 modes per processor core we obtain the desired efficiency of $\geq 65\%$ over a range of system sizes, at a cost of ≈ 3 seconds per full Runge-Kutta time-step. Note that the results of those code-performance study were first presented in Sakai and Uhlmann (2016).

2.3 Discretization strategies

2.3.1 Spatial discretization

We employ a collocated grid arrangement in the physical space, and the cross-sectional grid point distributions are based on the CGL points, viz.

$$y_j = \cos\left(\frac{\pi j}{M_y}\right) + 1 \quad \forall 0 \leq j \leq M_y, \qquad z_k = A\cos\left(\frac{\pi k}{M_z}\right) \quad \forall 0 \leq k \leq M_z\,. \tag{2.1}$$

nproc	walltime/time-step[sec]	speed-up	para. efficiency[%]
1	427.76	1.00	100
5	99.19	4.31	86
25	22.73	18.82	75.28
41	15.54	27.53	67.15
205	3.12	137.10	66.89
1025	2.21	193.56	19

Table 2.1: Strong scaling tests with the modal resolution of $M_x = 3075$, $M_y = 33$, $M_z = 1025$. This table is from Sakai and Uhlmann (2016).

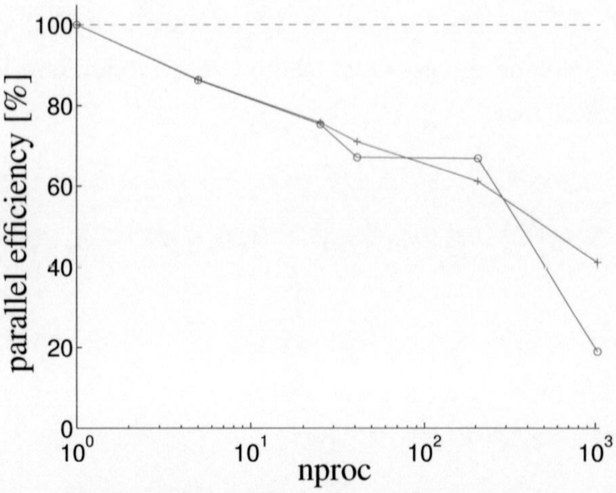

Figure 2.4: Parallel efficiency for two numbers of Chebyshev modes in y-direction: $M_y = 33$, ∘; $M_y = 67$, +. The number of modes in the x-/z-direction are fixed at $M_x = 3075$ and $M_z = 1025$ respectively. The runs were performed on Cray XC40 HORNET at HLRS. This figure is from Sakai and Uhlmann (2016).

where A is aspect ratio. The definitions of M_x, M_y, M_z will be mentioned shortly. Note that all computations involving the spectral transformations in (y,z) directions are performed on the standard CGL points by shifting and re-scaling the above cross-sectional points into $[-1,+1] \times [-1,+1]$ plane (i.e. $\tilde{y}_j = y_j - 1$, $\tilde{z}_k = z_k/A$, where $-1 \leq y(z) \leq +1$. The tilde symbol will be omitted hereafter for the sake of readability). In the streamwise direction, the following equidistant point distribution is used:

$$x_i = i\frac{L_x}{M_x} \quad \forall 0 \leq i \leq M_x - 1 \quad . \tag{2.2}$$

The flow variables are expanded in terms of truncated Fourier series in the streamwise direction and Chebyshev polynomials ($T_n($ y or $z$$)$) in the two cross-sectional directions, viz.

$$\psi(\boldsymbol{x}) = \sum_{i=0}^{M_x} \sum_{j=0}^{M_y} \sum_{k=0}^{M_z} \tilde{\psi}_{ijk} T_j(y) T_k(z) e^{2\pi I i x / L_x} , \tag{2.3}$$

where M_x, M_y and M_z are the upper limits of the truncated Fourier and Chebyshev modes in (x,y,z) directions respectively; $I = \sqrt{-1}$ is imaginary number; ψ represents any of the four variables u, v, w, p; and $\tilde{\psi}$ is the corresponding coefficient.

Additionally, we use the following notation for the variables in (Fourier) spectral space only in the streamwise direction:

$$\psi(\boldsymbol{x}) = \sum_{i=0}^{M_x} \hat{\psi}_i(y,z)\, e^{2\pi I i x/L_x} . \tag{2.4}$$

2.3.2 Temporal integration

We employ an *incremental-pressure projection method* (cf. Brown et al. (2001)) for splitting the governing equations (1.2) into two fractional steps. Using a semi-implicit scheme for the viscous terms and a three-step low-storage Runge-Kutta method with explicit non-linear terms, the semi-discrete system can be written as:

$$\frac{\boldsymbol{u}^* - \boldsymbol{u}^{k-1}}{\Delta t} = -\gamma_k \left[(\boldsymbol{u}\cdot\nabla)\boldsymbol{u}\right]^{k-1} - \zeta_k \left[(\boldsymbol{u}\cdot\nabla)\boldsymbol{u}\right]^{k-2} - 2\alpha_k \nabla p^{k-1} + \alpha_k \nu \nabla^2 \left(\boldsymbol{u}^* + \boldsymbol{u}^{k-1}\right) \tag{2.5a}$$

$$\nabla^2 \phi^k = \frac{\nabla\cdot\boldsymbol{u}^*}{2\alpha_k \Delta t} \tag{2.5b}$$

$$\boldsymbol{u}^k = \boldsymbol{u}^* - 2\alpha_k \Delta t \nabla \phi^k \tag{2.5c}$$

$$p^k = p^{k-1} + \phi^k - \alpha_k \Delta t\, \nu \nabla^2 \phi^k \tag{2.5d}$$

where $k=1,2,3$ is the Runge-Kutta step count (with $k=0$ being equivalent to the previous time-step and $k=3$ being the next time-step), $\boldsymbol{u}^*$ the predicted intermediate velocity, and the intermediate variable ϕ is sometimes referred to as *"pseudo pressure"*, and finally Δt is the time-step size for temporal integration. Note that the intermediate velocity $\boldsymbol{u}^*$ is not divergence-free, until being project onto the divergence-free space in the step (2.5c). Note also that all gradient evaluations are done in spectral space, whereas the non-linear terms are computed in the physical space.
The following set of coefficients, first employed by Rai and Moin (1991), leads to overall second-order temporal accuracy for both velocity and pressure in the interior of the simulation domain:

$$\alpha_k = \left\{\frac{4}{15}, \frac{1}{15}, \frac{1}{6}\right\}, \quad \gamma_k = \left\{\frac{8}{15}, \frac{5}{12}, \frac{3}{4}\right\}, \quad \zeta_k = \left\{0, -\frac{17}{60}, -\frac{5}{12}\right\}. \tag{2.6}$$

Moreover, the use of the fractional step technique introduces *slip error* (also sometimes referred to as *slip velocity*) into the solutions, whose magnitude is in an order of $\mathcal{O}(\Delta t^2 \nu)$.
The above formulations imply that the Helmholtz equations for the predicted velocity components (2.5a) and the Poisson equation for the pseudo-pressure (2.5b) need to be solved. In the velocity computations, no-slip boundary condition is imposed on the duct side/bottom-walls, whereas free-slip boundary condition is enforced on the free-slip plane. Conversely in the pseudo-pressure computations, zero-homogeneous Neumann boundary condition is employed. Those equations are solved in the physical space for each streamwise wavenumber, and the details of the employed technique will be explained in the following section.

Finally, a technique used to impose a prescribed constant mass-flow rate should be explained. It is realised by introducing an unknown constant pressure gradient P^* in Equation 2.5a, only for the 0th-Fourier mode of the streamwise predicted velocity $\hat{u}_0^*(y,z)$, viz.

$$\left[\nabla^2 - \frac{1}{\alpha_k \nu \Delta t}\right] \hat{u}_0^*(y,z) = \hat{F}_0(u^{k-1}, u^{k-2}, p^{k-1}) + P^* , \tag{2.7}$$

where $\hat{F}_0$ contains all the contributions from the explicit terms in Equation 2.5a to the 0th-Fourier mode.

The solution of the problem is decomposed as a linear combination of two components, namely $\hat{u}_0^I$ and $\hat{u}_0^{II}$. Whilst the former is the solution of Equation 2.5a with $P^* = 0$, the latter is the solution of the following equation:

$$\left[\nabla^2 - \frac{1}{\alpha_k \nu \Delta t}\right] \hat{u}_0^*(y,z) = 1 . \tag{2.8}$$

Then the final solution $\hat{u}_0^*(y,z)$ can be obtained as:

$$\hat{u}_0^*(y,z) = \hat{u}_0^I + P^* \hat{u}_0^{II} , \tag{2.9}$$

with P^*, which is determined by:

$$Q = \int_{\Omega_{cs}} \hat{u}_0^I \, dydz + P^* \int_{\Omega_{cs}} \hat{u}_0^{II} \, dydz , \tag{2.10}$$

where Q is the prescribed mass-flow rate through the duct cross-section Ω_{cs}.

2.4 Helmholtz solver

In this final part of the current chapter, we shall explain the details of the fast diagonalised method employed to solve the Helmholtz and Poisson problems discussed in the previous section.

The fast diagonalised technique is of Haldenwang and Labrosse (1984), and solves the Helmholtz and Poisson problems defined on the standard CGL grid points in both cross-sectional directions, where the field values are in the physical space in those two directions, whilst they are in the Fourier spectral space in the streamwise direction (cf. Equation 2.4). Either Dirichlet or Neumann boundary condition is imposed on each domain boundary.

The problems of consideration can be written in the following general form:

$$\nabla_\perp^2 \hat{u}(y,z) - c\hat{u}(y,z) = \hat{f}(y,z) , \tag{2.11}$$

where $\nabla_\perp^2 \equiv (\frac{\partial^2}{\partial y^2} + \frac{\partial^2}{\partial z^2})$, $\hat{u}(y,z)$ is the problem unknown, c is a constant, and $\hat{f}$ is a sum of the terms without $\hat{u}(y,z)$.

Subsequently, we consider a derivative matrix based on the collocated CGL points $y_j = \cos(\pi j/N_y)$ with $j = 0 \ldots N_y$ as:

$$\boldsymbol{D}_{l,m} = \begin{cases} \frac{c_l}{c_m}\frac{(-1)^{l+m}}{y_l - y_m} & l \neq m \\ \frac{-y_m}{2(1-y_m^2)} & 1 \leq l = m \leq N_y - 1 \\ \frac{2N_y^2+1}{6} & l = m = 0 \\ -\frac{2N_y^2+1}{6} & l = m = N_y \end{cases} \tag{2.12}$$

where $c_j = 1$ for $j = 1 \ldots N_y - 1$, and $c_0 = c_{N_y} = 2$ (cf. Canuto et al. (2006)). Note that we will refer to this $(N_y + 1) \times (N_y + 1)$ derivative matrix in y-direction as $\boldsymbol{D}_y$, and an equivalent $(N_z + 1) \times (N_z + 1)$ matrix in z-direction as D_z for $z_k = \cos(\pi k/N_z)$ with $k = 0 \ldots N_z$. In the following, the matrices and constants corresponding to the derivative matrix in each direction will be also indicated by the (y,z) subscripts. On the contrary, the properties that are common in both directions will be indicated without the subscripts (e.g. N instead of N_y).
Consequently, the discrete values $\hat{u}(y_j, z_k)$ and $\hat{f}(y_j, z_k)$ are stored in $(N_y + 1) \times (N_z + 1)$ matrices ($\boldsymbol{U}$, $\boldsymbol{F}$).
As a consequence, a discrete two-dimensional Helmholtz operator can be constructed as follows:

$$\left(\frac{\partial^2}{\partial y^2} + \frac{\partial^2}{\partial z^2} - c\right)\hat{u}(y,z) \simeq \boldsymbol{D}_y^2\boldsymbol{U} + \boldsymbol{U}(\boldsymbol{D}_z^2)^T - c\boldsymbol{U} \tag{2.13}$$

where $(\cdot)^T$ denotes a transposed matrix. The overall system of equations then becomes:

$$\boldsymbol{D}_y^2\boldsymbol{U} + \boldsymbol{U}(\boldsymbol{D}_z^2)^T - c\boldsymbol{U} = \boldsymbol{F} \,. \tag{2.14}$$

It turns out that, even if not symmetric, those $\boldsymbol{D}^2$ matrices have real-value eigensystems. Consequently, both matrices can be diagonalised as:

$$\boldsymbol{\Lambda} = \boldsymbol{P}^{-1}\boldsymbol{D}^2\boldsymbol{P} \;\rightarrow\; \boldsymbol{D}^2 = \boldsymbol{P}\boldsymbol{\Lambda}\boldsymbol{P}^{-1} \tag{2.15}$$

where $\boldsymbol{\Lambda}$ and $\boldsymbol{P}$ are the eigenvector and the eigenvalue matrix of a $\boldsymbol{D}^2$ matrix respectively. Furthermore, the same property holds for the transposed system, and the corresponding diagonalised system can be written as:

$$\boldsymbol{\Lambda} = \boldsymbol{P}^T(\boldsymbol{D}^2)^T\boldsymbol{P}^{-T} \;\rightarrow\; (\boldsymbol{D}^2)^T = \boldsymbol{P}^{-T}\boldsymbol{\Lambda}\boldsymbol{P}^T \tag{2.16}$$

where $(\cdot)^{-T}$ indicates the transposed of the inverse of a matrix.
Finally, above system of equations can be combined into one system of equations with a combination of boundary conditions of our choice. Moreover, the resulting system can be expressed in a condensed form with $(N-1) \times (N-1)$ matrices which will be denoted by the $\widetilde{(\cdot)}$ notation, by taking into account the selected boundary conditions. The details of the matrix condensation process will follow shortly. For that purpose, it is therefore important to note that the same real-eigensystem property holds for the following formulation.
The general form of such condensed system can be expressed as:

$$\widetilde{\boldsymbol{\Lambda}}_y\widetilde{\boldsymbol{P}}_y^{-1}\widetilde{\boldsymbol{U}}\widetilde{\boldsymbol{P}}_z^{-T} + \widetilde{\boldsymbol{P}}_y^{-1}\widetilde{\boldsymbol{U}}\widetilde{\boldsymbol{P}}_z^{-T}\widetilde{\boldsymbol{\Lambda}}_z - c\widetilde{\boldsymbol{P}}_y^{-1}\widetilde{\boldsymbol{U}}\widetilde{\boldsymbol{P}}_z^{-T} = \widetilde{\boldsymbol{P}}_y^{-1}\widetilde{\boldsymbol{F}}\widetilde{\boldsymbol{P}}_z^{-T} \tag{2.17}$$

where $\widetilde{U}$ and $\widetilde{F}$ are the condensed matrices containing the internal values of U and F, (i.e. $\widetilde{F}_{j,k} = f(y_j, z_k)$, $j = 1 \ldots N_y - 1$, $k = 1 \ldots N_z - 1$), whereas $\widetilde{P}$ and $\widetilde{\Lambda}$ are the eigenvector and eigenvalue matrices of $\widetilde{D^2}$. If the following transformations:

$$\widehat{U} = \widetilde{P}_y \widetilde{U} \widetilde{P}_z^T \quad \text{and} \quad \widehat{F} = \widetilde{P}_y^{-1} \widetilde{F} \widetilde{P}_z^{-T} \tag{2.18}$$

are introduced, then the linear system (2.17) results to be fully diagonal:

$$\widetilde{\Lambda}_y \widehat{U} + \widehat{U} \widetilde{\Lambda}_z - c\widehat{U} = \widehat{F} \tag{2.19}$$

and can be solved as:

$$\widehat{U}_{j,k} = \frac{\widehat{F}_{j,k}}{(\lambda_j + \lambda_k - c)} \tag{2.20}$$

The physical value of the solution is finally obtained applying the transform (2.18), together with the boundary values.

In the following, we will explain the aforementioned matrix condensation process and the related boundary condition treatments. The case of homogeneous Dirichlet boundary condition will be discussed first, which corresponds to the closed duct velocity computation. It will be then followed by the homogeneous Neumann case, which is the method applied to compute the pseudo-pressure for both closed and open duct flow. Finally, the Dirichlet-Neumann mixed-boundary case (i.e. the open duct velocity) will conclude the current chapter.

2.4.1 Homogeneous Dirichlet boundary problem

In the case of homogeneous Dirichlet problem, it is sufficient to eliminate the first and last rows and columns from the D^2 matrices. The boundary values are then reconstructed from the prescribed boundary condition.

2.4.2 Homogeneous Neumann boundary problem

Here we consider a case of homogeneous zero Neumann problems. The gradient conditions on the bottom ($y = -1$) and top ($y = +1$) domain boundaries at the location k are, respectively:

$$(D_y)_{0,0} U_{0,k} + \sum_{l=1}^{N_y-1} (D_y)_{0,l} U_{l,k} + (D_y)_{0,N_y} U_{N_y,k} = 0 \tag{2.21}$$

$$(D_y)_{N_y,0} U_{0,k} + \sum_{l=1}^{N_y-1} (D_y)_{N_y,l} U_{l,k} + (D_y)_{N_y,N_y} U_{N_y,k} = 0 \tag{2.22}$$

where $(D_y)_{l,m}$ are the (l,m) entries of D_y. In a similar manner, we can write the Neumann condition at the left ($z = -1$) and right ($z = +1$) boundaries at the location j:

$$(D_z)_{0,0} U_{j,0} + \sum_{l=1}^{N_z-1} (D_z)_{0,l} U_{j,l} + (D_z)_{0,N_z} U_{j,N_z} = 0 \tag{2.23}$$

$$(\boldsymbol{D}_z)_{N_z,0}\boldsymbol{U}_{j,0} + \sum_{l=1}^{N_z-1} (\boldsymbol{D}_z)_{N_z,l}\boldsymbol{U}_{j,l} + (\boldsymbol{D}_z)_{N_z,N_z}\boldsymbol{U}_{j,N_z} = 0 \tag{2.24}$$

Consequently, Equations (2.21, 2.22) can be solved for $\boldsymbol{U}_{0,k}$, $\boldsymbol{U}_{N_y,k}$ and Equations (2.23, 2.24) can be solved for $\boldsymbol{U}_{j,0}$, $\boldsymbol{U}_{j,N_z}$.
The obtained solution can be substituted in the system (2.14) to eliminate the boundary values. Operating in this way the (l,m) entries of the condensed form of the matrix $\boldsymbol{D}^2$ are:

$$\widetilde{\boldsymbol{D}^2}_{l,m} = a_{l,m} + \boldsymbol{D}^2_{l+1,m+1} + b_{l,m} \quad l = 1..N-1,\ m = 1..N-1 \tag{2.25}$$

with

$$a_{l,m} = \frac{\boldsymbol{D}^2_{l+1,0}\ (\boldsymbol{D}_{0,N}\ \boldsymbol{D}_{N,m} - \boldsymbol{D}_{N,N}\ \boldsymbol{D}_{0,m})}{\boldsymbol{D}_{0,0}\ \boldsymbol{D}_{N,N} - \boldsymbol{D}_{0,N}\ \boldsymbol{D}_{N,0}} \quad l = 1..N-1,\ m = 1..N-1 \tag{2.26}$$

and

$$b_{l,m} = \frac{\boldsymbol{D}^2_{l+1,N}\ (\boldsymbol{D}_{N,0}\ \boldsymbol{D}_{0,m} - \boldsymbol{D}_{0,0}\ \boldsymbol{D}_{N,m})}{\boldsymbol{D}_{0,0}\ \boldsymbol{D}_{N,N} - \boldsymbol{D}_{0,N}\ \boldsymbol{D}_{N,0}} \quad l = 1..N-1,\ m = 1..N-1 \tag{2.27}$$

Once the condensed matrices $\widetilde{\boldsymbol{D}^2}$ have been established, the discrete solution of the internal points can be obtained by using Equation 2.18–2.20.
To complete the solution procedure, we need to reconstruct the values of $\boldsymbol{U}$ on the boundaries. To this end, we use Equation 2.21–2.24, for all the boundary values except for the four corner points where a special treatment needs to be used. Here we can just use four unknown values to fix a homogeneous Neumann condition. Since the normal direction in a corner is not defined, we choose to impose the condition $\nabla u \cdot \boldsymbol{n} = 0$, using $\vec{n}$ as the $\pi/4$ direction oriented toward the inside of the square. This condition applied to the four corners lead to a 4×4 linear system of equations:

$$\begin{aligned} &((\boldsymbol{D}_y)_{0,0} + (\boldsymbol{D}_z)_{0,0})\boldsymbol{U}_{0,0} + \boldsymbol{D}_{0,N_y}\boldsymbol{U}_{N_y,0} + \boldsymbol{D}_{0,N_z}\boldsymbol{U}_{0,N_z} = \\ &\qquad - \sum_{l=1}^{N_y-1} (\boldsymbol{D}_y)_{0,l}\boldsymbol{U}_{l,0} - \sum_{m=1}^{N_z-1} (\boldsymbol{D}_z)_{0,m}\boldsymbol{U}_{0,m}; \end{aligned} \tag{2.28}$$

$$\begin{aligned} &-(\boldsymbol{D}_y)_{N_y,0}\boldsymbol{U}_{0,0} + ((\boldsymbol{D}_z)_{0,0} - (\boldsymbol{D}_y)_{N_y,N_y})\boldsymbol{U}_{N_y,0} + (\boldsymbol{D}_z)_{0,N_z}\boldsymbol{U}_{N_y,N_z} = \\ &\qquad \sum_{l=1,N_y-1} (\boldsymbol{D}_y)_{N_y,l}\boldsymbol{U}_{l,0} - \sum_{m=1}^{N_z-1} (\boldsymbol{D}_z)_{0,m}\boldsymbol{U}_{N_y,m}; \end{aligned} \tag{2.29}$$

$$\begin{aligned} &((\boldsymbol{D}_y)_{N_y,N_y} + (\boldsymbol{D}_z)_{N_z,N_z})\boldsymbol{U}_{N_y,N_z} + (\boldsymbol{D}_y)_{N_y,0}\boldsymbol{U}_{0,N_z} + (\boldsymbol{D}_z)_{N_z,0}\boldsymbol{U}_{N_y,0} = \\ &\qquad - \sum_{l=1}^{N_y-1} (\boldsymbol{D}_y)_{N_y,l}\boldsymbol{U}_{l,N_z} - \sum_{m=1}^{N_z-1} (\boldsymbol{D}_z)_{N_z,m}\boldsymbol{U}_{N_y,m}; \end{aligned} \tag{2.30}$$

$$-(\boldsymbol{D}_z)_{N_z,0}\boldsymbol{U}_{0,0} + ((\boldsymbol{D}_y)_{0,0} - (\boldsymbol{D}_z)_{N_z,N_z})\boldsymbol{U}_{0,N_z} + (\boldsymbol{D}_y)_{0,N_y}\boldsymbol{U}_{N_y,N_z} = - \sum_{l=1}^{N_y-1} (\boldsymbol{D}_y)_{0,l}\boldsymbol{U}_{l,N_z} \sum_{m=1}^{N_z-1} (\boldsymbol{D}_z)_{N_z,m}\boldsymbol{U}_{0,m} \, . \tag{2.31}$$

These equations are the discrete counterparts of the conditions:

$$\frac{\partial u}{\partial y} + \frac{\partial u}{\partial z} = 0 \quad \text{at} \quad y = z = -1 \, , \tag{2.32}$$

$$-\frac{\partial u}{\partial y} + \frac{\partial u}{\partial z} = 0 \quad \text{at} \quad y = 1, \, z = -1 \, , \tag{2.33}$$

$$-\frac{\partial u}{\partial y} - \frac{\partial u}{\partial z} = 0 \quad \text{at} \quad y = 1, \, z = 1 \, , \tag{2.34}$$

$$\frac{\partial u}{\partial y} - \frac{\partial u}{\partial z} = 0 \quad \text{at} \quad y = -1, \, z = 1 \, . \tag{2.35}$$

The linear system can be solved in the four corner values, once the value of the unknown on the four edges has been determined.

Finally, we describe the procedure that has been employed to eliminate the singularity from the Neumann problem with $c = 0$. If we look at Equation (2.20), it is clear that to one of the variable will correspond the equation:

$$(\lambda_{\hat{j}} + \lambda_{\hat{k}})\widehat{\boldsymbol{U}}_{\hat{j},\hat{k}} = \widehat{\boldsymbol{F}}_{\hat{j},\hat{k}} \, , \tag{2.36}$$

where $\hat{j}$ and $\hat{k}$ the index of the zero eigenvalue of matrix $\widetilde{\boldsymbol{D}_y^2}$ and $\widetilde{\boldsymbol{D}_z^2}$ for the Neumann case. No value of $\widehat{\boldsymbol{U}}_{\hat{j},\hat{k}}$ will verify this equation. To eliminate this singular behaviour, we eliminate equation $\hat{j},\hat{k}$ from the system by substituting the condition $\boldsymbol{U}_{\hat{j}+1,\hat{k}+1} = \Phi$, where Φ is a prescribed fixed value. The transform step (2.18) for this particular node can be written as:

$$\Phi = \boldsymbol{U}_{\hat{j}+1,\hat{k}+1} = \sum_{l=1}^{N_y-1} (\widehat{\boldsymbol{P}_y})_{\hat{j},l} \left(\sum_{m=1}^{N_z-1} (\widehat{\boldsymbol{P}_z})_{\hat{k},m} \widehat{\boldsymbol{U}}_{l,m} \right) \, . \tag{2.37}$$

Once all the values, except the critical one, have been determined using (2.36), we can find the value of $\widehat{\boldsymbol{U}}_{\hat{j},\hat{k}}$ that assign to $\boldsymbol{U}_{\hat{j}+1,\hat{k}+1}$ with the value Φ:

$$\widehat{\boldsymbol{U}}_{\hat{j},\hat{k}} = \frac{1}{(\widehat{\boldsymbol{P}_y})_{\hat{j},\hat{j}}(\widehat{\boldsymbol{P}_z})_{\hat{k},\hat{k}}} \left(\Phi - \sum_{l=1\ l\neq\hat{j}}^{N_y-1} (\widehat{\boldsymbol{P}_y})_{\hat{j},l} \left(\sum_{m=1\ m\neq\hat{k}}^{N_z-1} (\widehat{\boldsymbol{P}_z})_{\hat{k},m} \widehat{\boldsymbol{U}}_{l,m} \right) \right) \, . \tag{2.38}$$

2.4.3 Homogeneous Dirichlet-Neumann mixed-boundary problem

We employ the zero homogeneous Neumann boundary condition at $j = N_y$ (i.e. Equation 2.22) and the rest will be zero homogeneous Dirichlet boundary condition. Then the equation can be solved for $\boldsymbol{U}_{N_y,k}$ as:

$$\boldsymbol{U}_{N_y,k} = \frac{-(\boldsymbol{D}_y)_{Ny,0}\boldsymbol{U}_{0,k} - \sum_{l=1}^{N_y-1}(\boldsymbol{D}_y)_{N_y,l}\boldsymbol{U}_{l,k}}{(\boldsymbol{D}_y)_{N_y,N_y}} , \tag{2.39}$$

where $\boldsymbol{U}_{0,k} = 0$ by the Dirichlet boundary condition. Consequently, Equation 2.14 can be reduced to $(N_y - 1) \times (N_z - 1)$ system of equations by substituting the above expression, viz.

$$\sum_{l=1}^{N_y-1} \left((\boldsymbol{D}_y^2)_{j,l} - \frac{(\boldsymbol{D}_y)_{N_y,l}}{(\boldsymbol{D}_y)_{N_y,N_y}} (\boldsymbol{D}_y^2)_{j,N_y} \right) \boldsymbol{U}_{l,k} + \sum_{m=1}^{N_z-1} (\boldsymbol{D}_z^2)_{k,m} \boldsymbol{U}_{j,m} - c\boldsymbol{U}_{j,k} = \boldsymbol{F}_{j,k} . \tag{2.40}$$

Finally, the condensed $\boldsymbol{D}_y^2$ matrix can be constructed as:

$$(\widetilde{\boldsymbol{D}_y^2})_{l,m} = (\boldsymbol{D}_y^2)_{l+1,m+1} - (\boldsymbol{D}_y^2)_{l+1,N_y} \frac{(\boldsymbol{D}_y)_{N_y,m+1}}{(\boldsymbol{D}_y)_{N_y,N_y}} \quad l = 1..N_y - 1, \; m = 1..N_y - 1 . \tag{2.41}$$

Chapter 3

Origin and Reynolds number dependence of secondary motions in open duct flow

3.1 Introduction

The main objective of this chapter is to determine the origin of the secondary motions in open duct turbulence from the viewpoint of the turbulent coherent structures and their dynamics.
In order to do so, the lower limit of sustainable Reynolds number will be determined first, where a scale separation of fluid motions does not exist and, therefore, the secondary motions are expected to be direct consequences of the coherent structures themselves. Under such condition, any movements of the coherent structures are severely constrained by the duct cross-sectional geometry, and their preferential locations should become noticeable within the time-averaged fields as a consequence. Identifying the key coherent structures in the marginally turbulent state is therefore expected to be relatively straightforward in spite of the chaotic nature of turbulent flow.
Subsequently, the flow Reynolds number will be increased gradually up to a point where the scale separation becomes significant. Investigations of the impact of Reynolds number upon the coherent structures should provide us further insight into the scaling properties of the open duct secondary flow. Based upon the discovered scaling properties supplemented by extensive instantaneous vortex eduction analysis, which is analogous to the ones performed to the closed duct counterpart previously (cf. Uhlmann et al. (2007), Pinelli et al. (2010), Sekimoto et al. (2011)), we shall attempt to describe the origin of the mean secondary flow patterns from the viewpoint of the coherent structure dynamics.
Moreover, possible Reynolds number dependence in other important phenomena such as velocity dip-phenomenon as well as integral measures of the mean secondary quantities will be discussed. The outcomes of the scaling investigations should serve as key ingredients to verify some of the empirical approximations employed frequently in engineering practice, many of which are based on classical experimental measurements with marginal resolutions. Furthermore, a new kind of coherent structure eduction technique, which takes into account the length-scales of individual

structures, will be employed to explain the dip-phenomenon scaling as a consequence of the preferential locations of large-scale structures.
Throughout this chapter, many comparative remarks will be made against the corresponding square closed duct flow. Therefore in order to isolate any possible influences, the cross-sectional aspect ratio is fixed at unity.

3.2 Simulation set-up

All simulations presented here satisfy the following resolution requirements: CFL number is below 0.3; Δx^+ is below 15.2; the number of Chebyshev mode ensures that the maximum grid spacings in the cross-sectional directions $\max\{\Delta y^+\}$ and $\max\{\Delta z^+\}$ are below 4.1. Note that due to the inflexible nature of the Chebyshev–Gauss–Labatto points, the near-boundary resolutions, including the one beside the free-slip plane where a thin boundary layer is known to form and needs to be resolved, are in general extremely high (i.e. $\min\{\Delta y^+\} \approx \min\{\Delta z^+\} = \mathscr{O}(10^{-2})$).
In order to demonstrate the adequateness of the above spatial resolution criteria, a series of grid-convergence test was performed for the baseline open duct case with $A = 1$ and $Re_b = 2205$, by refining or coarsening the number of modes in all three directions by a factor of ≈ 1.5, which were initially set at $[M_x, M_y, M_z] = [256, 97, 193]$. Note that the default spatial resolutions expressed in the wall units are $\Delta x^+ = 14.8$ and $\max\{\Delta y^+\} \approx \max\{\Delta z^+\} = 2.46$. The results (summarised in Appendix A) show that the coarser resolution is insufficient to resolve the flow near the free-slip plane, while the baseline and refined resolutions produced indistinguishable results.
Sufficiency of the streamwise extension $L_x/H = 8\pi$ was also examined in terms of two-point correlation of the normal components of Reynolds stresses (again shown in Appendix A), and the results showed sufficient decorrelation at all four representative locations of the cross-sectional plane. Note that the decorrelation within the mixed-boundary corner requires the largest streamwise separation, indicating that a numerical domain for open duct flow needs to be significantly longer than the ones used in the previous closed duct simulations (e.g. $L_x/h = 4\pi$ in Pinelli et al. (2010), Zhang et al. (2015)).
In total 15 open duct simulations were run until their statistical convergence for this study, by setting a constant aspect ratio of unity and varying the bulk Reynolds number from 1450 to 7000. In addition, a new set of closed duct simulations with Re_b ranging from 4000 to 7000 ($Re_\tau = 251 - 411$) with $L_x/h = 4\pi$ were performed for comparison purpose. Their statistics were accumulated over at least 4121 H/u_b in order to achieve satisfactory convergence. Numerical parameters of those simulations are summarised in Table 3.1 and 3.2 for the open and closed duct simulations respectively.

3.3 Results

3.3.1 Minimal Reynolds number

Through the simulation campaign, the minimum sustainable Reynolds number for the current configuration was found at $Re_b = 1450$, which corresponds to $Re_\tau \approx 102$.

Re_b	Re_τ	$N_x \times N_y \times N_z$	Δx^+	$\max\{\Delta y^+, \Delta z^+\}$	$t_{stat} u_b / H$
1450	102	$256 \times 97 \times 193$	10.2	1.69	9183
1480	103	$256 \times 97 \times 193$	10.1	1.68	5248
1500	104	$256 \times 97 \times 193$	10.2	1.69	11807
1550	107	$256 \times 97 \times 193$	10.4	1.74	9180
1800	125	$256 \times 97 \times 193$	12.3	2.05	11449
2205	150	$256 \times 97 \times 193$	14.8	2.46	12940
3000	197	$384 \times 97 \times 193$	12.9	3.23	9082
3500	226	$384 \times 97 \times 193$	14.8	3.70	14826
4000	254	$512 \times 129 \times 257$	12.5	3.12	9016
4500	282	$512 \times 129 \times 257$	13.8	3.45	10306
5000	309	$512 \times 129 \times 257$	15.2	3.80	8005
5500	336	$768 \times 193 \times 385$	11.0	2.75	4121
6000	363	$768 \times 193 \times 385$	11.9	3.00	7282
6500	388	$768 \times 193 \times 385$	12.7	3.20	5064
7000	415	$768 \times 193 \times 385$	13.6	3.40	4765

Table 3.1: Parameters in open duct simulations. All runs are with $A = 1$.

Re_b	Re_τ	$M_x \times M_y \times M_z$	Δx^+	$\max\{\Delta y^+, \Delta z^+\}$	$t_{stat} u_b / h$
4000	251	$512 \times 257 \times 257$	6.2	3.1	8875
4500	278	$512 \times 257 \times 257$	6.8	3.4	8326
5000	306	$512 \times 257 \times 257$	7.5	3.7	6501
5500	332	$512 \times 257 \times 257$	8.2	4.1	6501
6000	359	$512 \times 385 \times 385$	8.8	2.9	12819
6500	385	$512 \times 385 \times 385$	9.5	3.2	7756
7000	411	$512 \times 385 \times 385$	10.9	3.4	6405

Table 3.2: Parameters in additional closed duct simulations. All runs are with $A = 1$.

The procedure and criteria to define the minimum sustaining Reynolds number is as of Uhlmann et al. (2007), by initialising the flow fields with a fully developed turbulence taken from a previous run at higher Reynolds number. In this fashion the Reynolds number has been decreased in steps of approximately 20-50 bulk units. The computations were run until relaminarisation, or the flow remained turbulent over 2000 bulk time units in which case the flow is marked as sustainable and the Reynolds number is further reduced. The existence of turbulent activities is assessed by monitoring the time-development of the turbulent kinetic energy that drops to zero in a case of relaminarisation. Note that no attempt of a statistical analysis of the life-time of turbulence (e.g. Hof et al. (2006)) has been made.

The critical value that was found in this study is significantly higher than the value found in the closed duct configuration which is at $Re_\tau \approx 77$ (Uhlmann et al., 2007) for the same aspect ratio at unity, which means their duct half-height h (and span) needs to be at least 77 in viscous units. The equivalent minimum value of H^+ for the current configuration is at 102, which seems considerably more restrictive by taking into account the fact that the considered open duct geometry is identical to the closed duct counterpart except for imposing the symmetry condition at the wall bi-sector.

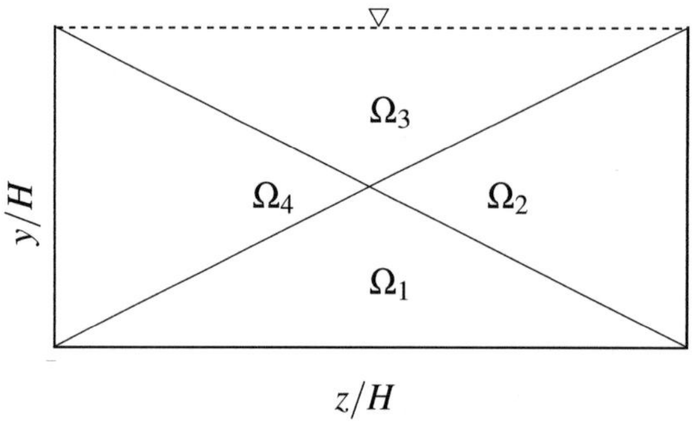

Figure 3.1: Schematic of triangular sub-domain Ω_i

The source of such discrepancy will be discussed in the following sections. Please also note that the streamwise-localised turbulent structures (i.e. "*puffs*") are known to be sustained at even lower Reynolds numbers in transient regime, if the streamwise domain extension is set to be significantly longer (e.g. $L_x = 40\pi h$ in Takeishi et al. (2015)).

Within the same study (Uhlmann et al. (2007)), they demonstrated the existence of a temporary-alternating four-vortex state with a $\pi/2$-rotational symmetry, forming the well-known eight-vortex pattern only in long-term statistics for marginal Reynolds numbers below $Re_b \approx 1250$. On the other hand, such clear switching of orientation has not been observed from the current open duct data even around the lowest Reynolds number achieved in this study, possibly due to the relatively high value ($Re_b = 1450 > 1250$).

In order to demonstrate the absence of the temporal alternation of the orientation, an indicator function analogous to the one employed in Uhlmann et al. (2007) is introduced here. The indicator function $S_i(t)$ is based on the time-evolution of the streamwise-averaged squared streamwise vorticity integrated over one of the four equiareal triangular regions on the duct cross-section:

$$S_i = \int\int_{\Omega_i} \langle \omega_x \rangle_x^2 dydz \,, \tag{3.1}$$

where the triangular regions Ω_i are shown in Figure 3.1 and defined as:

$$\begin{cases} \Omega_1 \, : \, \{(y,z)|(2y-H)/H < z/(AH) \cap (2y-H)/H < -z/(AH)\}, \\ \Omega_2 \, : \, \{(y,z)|(2y-H)/H < z/(AH) \cap (2y-H)/H > -z/(AH)\}, \\ \Omega_3 \, : \, \{(y,z)|(2y-H)/H > z/(AH) \cap (2y-H)/H > -z/(AH)\}, \\ \Omega_4 \, : \, \{(y,z)|(2y-H)/H > z/(AH) \cap (2y-H)/H < -z/(AH)\}. \end{cases} \tag{3.2}$$

Consequently, we introduce the following non-dimensional indicator function:

$$I(t) = \frac{S_1 + S_3 - S_2 - S_4}{S_1 + S_2 + S_3 + S_4} \tag{3.3}$$

that is bounded by $-1 \leq I_1 \leq 1$. The outcome of the indicator function is designed to be zero when the streamwise vorticity is equipartitioned between those two triangular areas associated with the walls at $y = [0, H]$ and the two associated with the walls at $z = \pm AH$. Note that due to the imposed heterogeneous boundary conditions, the indicator is not expected strictly to be at zero conversely to the eight-vortex state in the square closed duct flow. In any case, we expect the time variation

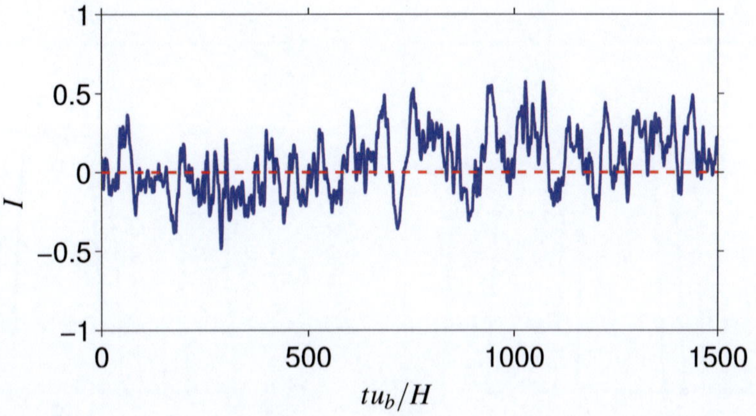

Figure 3.2: Temporal evolution of the indicator function I of the open duct simulation with $Re_b = 1450$ and $A = 1$. Red dashed line indicates $I = 0$.

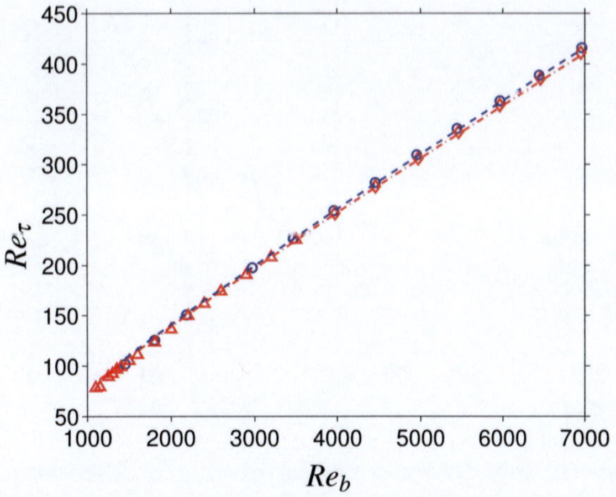

Figure 3.3: Friction Reynolds number Re_τ vs. bulk Reynolds number Re_b. Symbols indicate: ○, open duct; △, closed duct. Blue dashed and red dash-dot line indicate predictions by the empirical formula of Jones (1976).

of the indicator value to show a relatively low amplitude and a high frequency of fluctuations in case of the absence of the noticeable change of the orientation.

On the contrary, if any noticeable temporal alternations between the side and bottom-walls exist, such as the four-vortex state observed in the marginal turbulence in the square closed duct flow, instantaneous concentration of the streamwise vorticity would be one-sided and the indicator function should yield step-like quasi-periodic transitions between two relatively high values, with a relatively long alternation period (cf. Figure 4(a) in Uhlmann et al. (2007)).

Figure 3.2 shows a time evolution of of the indicator function I from the current open duct simulation with $Re_b = 1450$, and it can be seen that there are no step-like transitions, but there are low-amplitude high-frequency fluctuations around zero that are similar to the fully-developed closed duct turbulence instead. The time-averaged value of I in the current simulation is at 0.0819.

Concerning the friction factor, the empirical formula by Jones (1976) was confirmed to hold also for the current flow, with a form $f^{-1/2} = 2\log_{10}(2.0208\ Re_b\ f^{1/2}) - 0.8$, where $f = 8u_\tau^2/u_b^2$, by taking into account the difference in the geometrical aspect ratio from the closed duct flow with the identical D_{H} (cf. Figure 3.3).

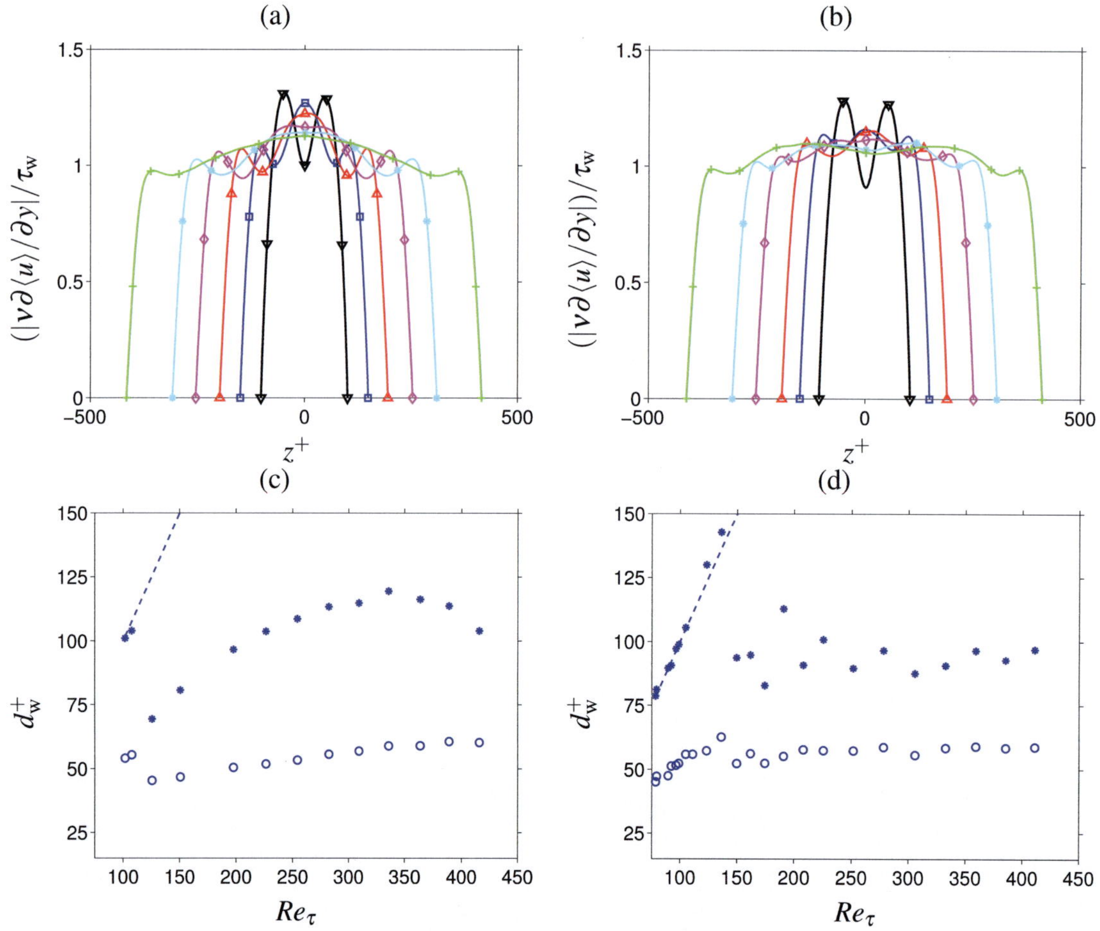

Figure 3.4: (a,b) Mean bottom-wall stress distributions of: (a) open and (b) closed ducts, normalised by the whole no-slip wall average value τ_w and presented as a function of z^+. Coloured symbols indicate the value of the bulk Reynolds numbers: $\bigtriangledown$, $Re_b = 1450$ ($Re_b = 1500$ in (b)); $\square$, $Re_b = 2205$; $\triangle$, $Re_b = 3000$; $\diamond$, $Re_b = 4000$; $*$, $Re_b = 5000$; $+$, $Re_b = 7000$. (c,d) Distance to the closest corner d_w^+, in wall units, of the location of the near-corner maxima and minima of bottom-wall shear as a function of Re_τ. (c) open and (d) closed ducts. Here symbols $\circ$ and $*$ represent the position of the shear maxima and minima respectively. The dashed line indicates duct mid-span in wall units H^+ as a function of Re_τ.

3.3.2 Mean streamwise structure

Here we associate the preferential locations of high- and low-velocity streaks with the local maxima and minima of the mean wall shear stress distribution, such as the ones shown in Figure 3.4(a,b) for open and closed duct flow for a range of representative Reynolds number respectively. From the stress profile along the duct bottom-wall shown in Figure 3.4(a), with a special emphasis upon the locations and number of their local minima and maxima, it is clearly visible that the lowest Reynolds number $Re_b = 1450$ case accommodates three velocity streaks above the bottom-wall edge in high-low-high order with the local wall stress minima preferentially located at the duct mid-span.

In fact for $Re_b < 1800$, the location of the low-velocity streak minima is geometrically constrained at the duct mid-span (cf. Figure 3.4(c)), indicating the argument of the geometrical upper bound of the habitable number of velocity streaks by Pinelli et al. (2010) holds also for the open duct bottom-wall region. Note that the argument is based on the fact that the average distance between same-signed streaks is in $\mathcal{O}(100^+)$ (Kim et al. (1987)). At $Re_b \approx 1800$, the bottom-wall edge should theoretically be able to accommodate five streaks in high-low-high-low-high order over ≈ 300 wall units. Indeed a sharp change in the low-velocity streak location at that Reynolds number in Figure 3.4(c) confirms such change of state.

As Reynolds number is further increased, signatures of the preferential velocity streak locations become progressively less obvious except for the high- and low-velocity streak pairs beside the corners. Pinelli et al. (2010) reported that the location of the first high-velocity streak maxima is consistently located at around 50 wall units away from the corners, that is again also consistent with our open duct data (cf. Figure 3.4(c)).

Alongside those similarities, differences between the open and closed duct statistics should be also noted, such as the elevated shear distribution over the duct mid-span region relative to the values in the corner region. The unique outer secondary flow motion transporting the high-momentum fluid from the free-slip plane towards the bottom-wall region is believed to be responsible for the higher stress distribution. Furthermore, the aforementioned high-velocity streak locations (i.e. '$\circ$' symbols in Figure 3.4(c)) have slight Reynolds number dependence although the variation can be argued that is diminishing at the higher end of the considered Reynolds number range. The low-velocity streak locations ('$*$' symbols in Figure 3.4(c)) show more noticeable Reynolds dependence: the distance from the side-wall first increases upto $Re_b \approx 5500$, then it decreases as Reynolds number increases. Note that those modifications to the statistics that cannot be explained by the small-scale near-wall dynamics, but possibly by their interactions with the large-scale mean secondary motions. Further investigations are required to quantify such interactions between different length-scales of motions.

On the side-walls, differences in comparison with the closed duct counterpart is much more visible, such as the fact that the first high-velocity streaks from the mixed-boundary corners are located directly on the free-slip plane (cf. Figure 3.5(a)).

This persistent existence of the free-slip plane high-velocity streaks is remarkable, in a sense that there are at least two high-velocity streaks on each side-wall edge even at the marginal limit of Reynolds numbers. Here, a clear contrast should be drawn for this nature on the open duct wall shear distribution against the half-cut view of the wall shear profile in the marginal closed duct turbulence, where only one pair of low- and high-velocity streaks can exist. Therefore, this qualitative influence should be interpreted as a signature of the free-slip, impermeable boundary condition imposed on the upper boundary of the cross-section, in contrast to the plane of the statistical symmetry in the closed duct flow.

Consequently, the existence of the additional high-velocity streaks can be used to explain the higher minimum Reynolds number found in this study from the viewpoint of the geometrical constrains upon the minimal unit of the turbulent velocity streak structures necessary for the flow to remain being turbulent. In other words, each side-wall edge of the minimal open duct flow needs to be longer in wall units in order to accommodate one additional velocity streak in comparison with the closed duct counterparts, hence requires higher Reynolds number.

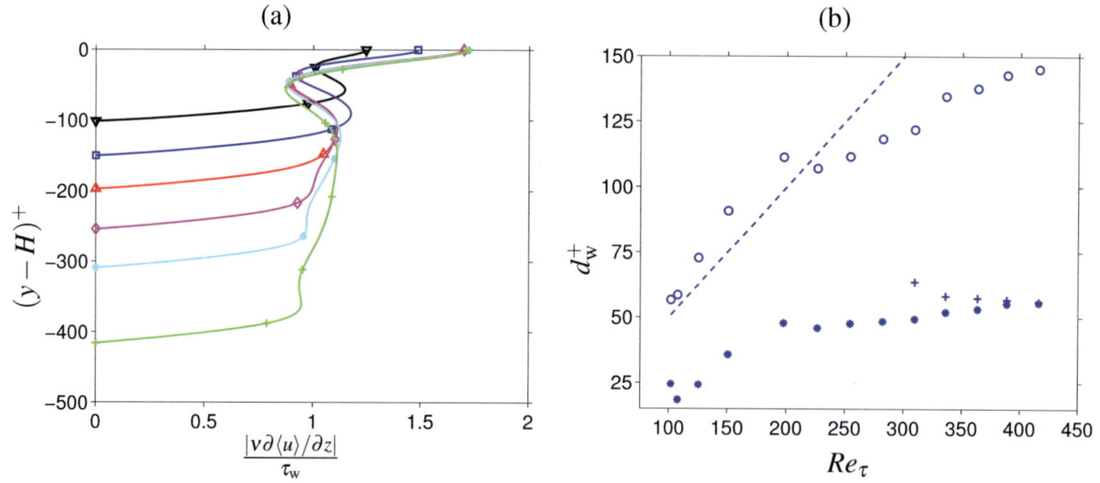

Figure 3.5: (a) Mean side-wall stress distribution normalised by the average over the whole no-slip wall τ_w. The origin is shifted to the mixed-boundary corner and the distance is presented in wall units. The coloured symbol code is as of Figure 3.4. (b) Distance to the prescribed corners d_w^+, in wall units, of the side-wall stress maxima and minima as a function of Re_τ. Here $*$ is the position of minima next to the free-slip plane, while $\circ$ represent the location of the maxima directly below. Additionally, $+$ symbol indicate the maxima forms next to the solid-solid corner for $Re_b \geq 5000$. The dashed line is duct half-height in wall units h^+ as a function of Re_τ.

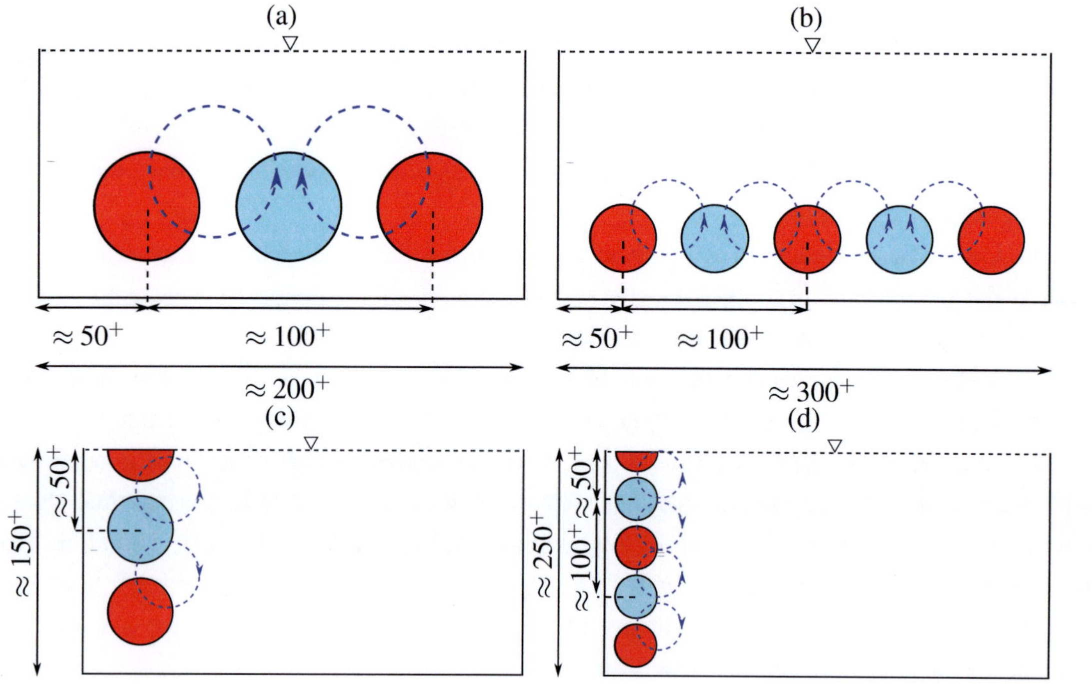

Figure 3.6: Schematics of (a,c) three-streak state; and (b,d) five-streak state, located on the open duct (a,b) bottom-wall edge; and (c,d) side-wall edge. For simplicity, only left side-wall structures are shown in (c,d). Red- and cyan-coloured filled circles represent high-velocity and low-velocity streaks respectively, whereas blue dashed lines with an arrow head represent near-wall vortices.

Furthermore, with increasing Reynolds number the location of the subsequent low-velocity streak moves towards and eventually settles down at about 50 wall units away from the mixed-boundary corners for $Re_b \geq 3000$ ($Re_\tau \geq 200$), in contrast to the bottom-wall low-velocity streaks that do not have such strict preferable location once the aforementioned geometrical constrain is removed for the elevated Reynolds numbers. Note that such persistent locations of the near-corner low-velocity streaks were previously reported by Grega et al. (1995) in their experimental studies. Also note that the concerned Reynolds number $Re_b = 3000$ ($Re_\tau = 200$) is where five velocity streaks start to fit on each side-wall edge, by taking into account the new finding that there is no gap necessary between the free-slip plane and the first high-velocity streaks from the plane.

Eventually at $Re_b = 5000$ ($Re_\tau = 309$), which corresponds to the theoretical lower limit of seven-streak state on the side-wall edges, an additional high-velocity streak starts to form close to the bottom-wall solid-solid corner. The corresponding distance from the bottom-wall boundary is again $d_w^+ \approx 50$ (cf. '+' symbols in Figure 3.5(b)), which coincides well with the preferential locations of the high-velocity streaks reside inside the solid-solid corners (cf. earlier discussion on the bottom-wall shear stress distribution). In fact, the high-velocity streaks responsible for the side-wall/ bottom-wall stress extrema next to the solid-solid corners need to be an identical structure, since those side-wall/bottom-wall edges share the same corner geometrically. Schematic diagrams summarising those different numbers of streak states along the bottom-wall and side-wall edges are separately presented in Figure 3.6.

3.3.3 Mean crossflow structure

3.3.3.1 Mean secondary flow

The unique mean secondary flow pattern of the open duct turbulence for a set of representative Reynolds numbers of $Re_b = [1450, 2205, 5000, 7000]$ can be seen in Figure 3.7 in terms of mean secondary flow streamfunction $\langle \psi \rangle$ contours.

Note that $\langle \psi \rangle$ is a solutions of the following Poisson's equation solved with zero-homogeneous Dirichlet boundary condition: $\nabla^2 \langle \psi \rangle = -\langle \omega_x \rangle$ with $\langle \omega_x \rangle$ that is the mean streamwise vorticity. There is one relatively small swirl inside the mixed-boundary corner, and a large opposite-signed roll circumventing the first one along the free-slip plane towards the duct mid-span then towards the duct core, which are commonly referred as inner and outer secondary flow cells. Additionally, another mean vortex cell exists in the lower part of the domain beside the solid-solid corner, rotating in the opposite to the rotational direction of the outer secondary flow cell.

It is important to observe for the lowest Reynolds number at $Re_b = 1450$ that the profile is qualitatively similar to the ones from the closed duct flow, having a pair of counter-rotating mean vortices that are in an almost equal size and symmetric about the solid-solid corner diagonals, but with one exception to have the additional inner secondary flow cell inside the mixed-boundary corners. This appearance of the extra mean vortices, due to the impermeability boundary condition imposed on the upper limit of the domain cross-section, corresponds to the existence of the additional high-velocity streaks inside the mixed-boundary corners as discussed in the previous section. Also it can be used to explain again the higher minimum Reynolds number limit from the viewpoint of the geometrical constraint imposed upon the turbulent coherent structures: the open duct cross-sectional dimensions in wall units need to be larger in order to confine those extra vortices.

Subsequently, the dimension of the outer secondary flow cell becomes significantly larger than the bottom-wall one at around $Re_b = 1800$, where it coincides with when the number of the bottom-wall streaks shifts from three to five (cf. Figure 3.6(a,b)). The dimensions of the mean secondary flow cells then become invariant except the one for the inner secondary flow cell whose diameter becomes continuously smaller as Reynolds number increases.

This Reynolds number dependence of the inner secondary flow cell size is rather surprising, since it indicates a separate scaling, based on the viscous length-scale δ_v in contrast to the other two which are scaling with the bulk length-scale H, simultaneously in the same domain. Note that the mean flow patterns in the square duct flow are shown to be Reynolds number independent by Pinelli et al. (2010) except some details deep inside the corners, indicating a predominant influence by the large-scale motions that are constrained by the duct geometry.

In order to quantify the scaling properties, the location of those streamfunction peaks are determined numerically by means of two-dimensional Newton's method from the tailored initial

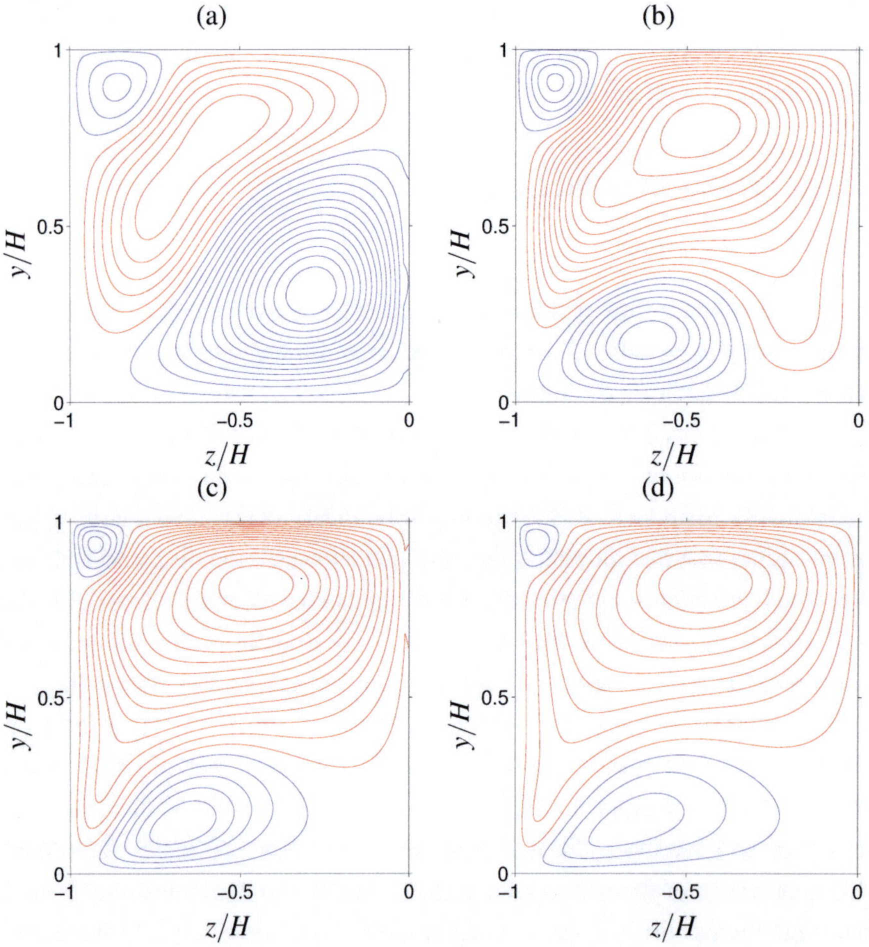

Figure 3.7: Streamfunction $\langle\psi\rangle$ contours of mean secondary flow computed using $\langle v\rangle$ and $\langle w\rangle$ with reflective symmetry about duct mid-span (with increment $[\max\langle\psi\rangle - \min\langle\psi\rangle]/20$). Red lines correspond to clockwise rotation and blue lines to counter-clockwise motion. (a) $Re_b = 1450$; (b) $Re_b = 2205$; (c) $Re = 5000$; (d) $Re_b = 7000$. Only left-half of the duct cross-section ($[0,H] \times [-AH,0]$) is shown.

guesses. The determined locations are then expressed as distances from the closest boundaries in y- and z-directions, and plotted against Reynolds number in Figure 3.8 for the outer secondary flow and the bottom-wall vortex cells, whereas Figure 3.9 shows the distances for the inner secondary flow cell. In both figures, those values are normalised separately by: (a) duct full-height H; and (b) viscous length-scale δ_v. It can be clearly seen that the peak locations of the outer secondary flow, as well as of the bottom-wall vortex cells show bulk-scaling properties after some initial transition takes place for $Re_b < 3000$. On the other hand, the inner secondary flow cell peak location scales better in the wall units for the same range of Reynolds number, ranging between 20^+ and 30^+ from the closest boundaries. Note that the range coincides well with the peak location of the root-mean-square of the streamwise vorticity fluctuation ($\sqrt{\langle \omega_x' \omega_x' \rangle}$) inside the buffer layer of the plane channel flow by Kim et al. (1987), indicating a significant link between the inner secondary flow and the buffer-layer quasi-streamwise vortices.

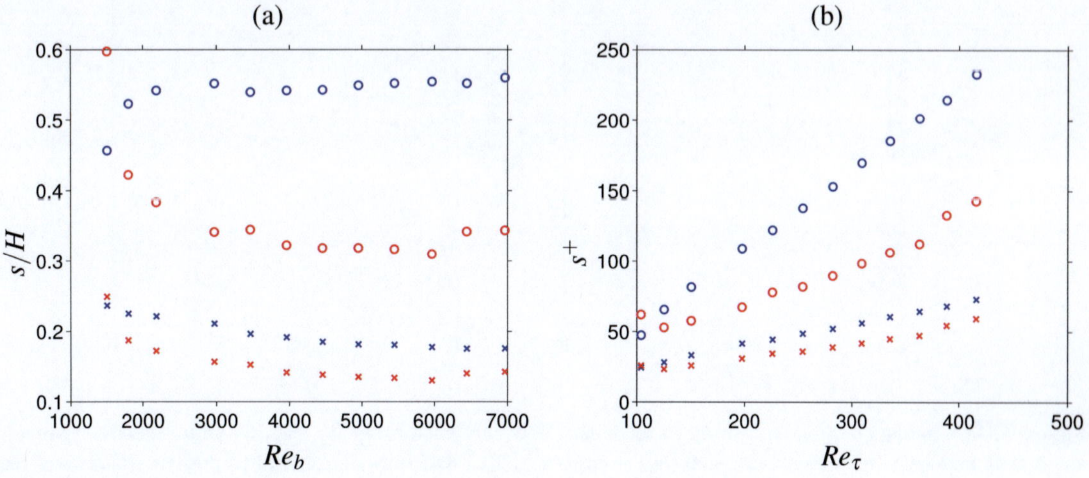

Figure 3.8: Positions of extrema of $\langle \psi \rangle (y,z)$: (a) external coordinates vs. Re_b and (b) wall coordinates vs. Re_τ. Colours: blue, outer secondary flow cell; red, bottom-wall cell. Symbols: $\circ$ distance to $z/H = -A$; $\times$ distance to $y/H = 1$ (blue symbols) and $y/H = 0$ (red symbols).

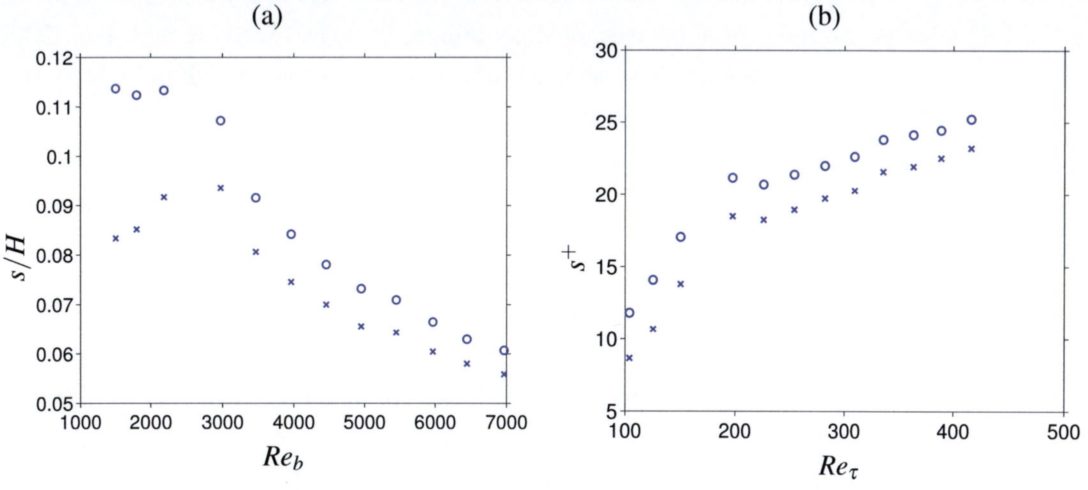

Figure 3.9: Positions of extrema of $\langle \psi \rangle (y,z)$ of inner secondary flow cell in: (a) external coordinates vs. Re_b and (b) wall coordinates vs. Re_τ. Symbols: $\circ$ distance to $z/H = -A$; $\times$ distance to $y/H = 1$.

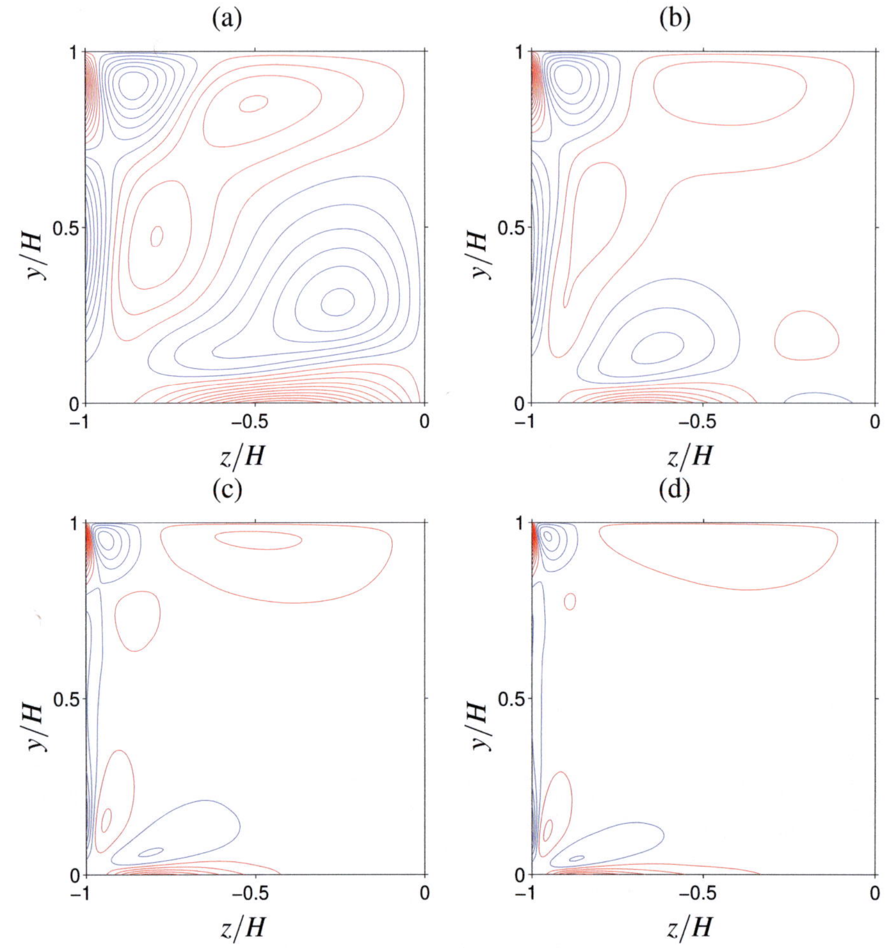

Figure 3.10: Contours of $\langle\omega_x\rangle(y,z)$ of mean secondary flow computed using $\langle v\rangle$ and $\langle w\rangle$ with reflective symmetry about duct mid-span (with increment $[\max\langle\omega_x\rangle - \min\langle\omega_x\rangle]/20$). Red lines correspond to positive values and blue lines to negative ones. (a) $Re_b = 1450$; (b) $Re_b = 2205$; (c) $Re = 5000$; (d) $Re_b = 7000$. Only left-half of the duct cross-section ($[0,H] \times [-AH,0]$) is shown.

Grega et al. (1995) reported that the mean diameter of the inner secondary flow cells is around $d^+ \approx 100$, whereas the result from the current study implies $d^+ \approx 60$. Later, Grega et al. (2002) showed that the inner secondary cell is located approximately at 75 and 35 wall units away from the free surface and the closest side-wall respectively. Note that our separate studies with variable aspect ratio (cf. Chapter 4) indicates that their narrow aspect ratio $A = 0.24$ is a likely cause of the discrepancy between their observations and the current results, including the enlarged dimension of the inner secondary cell.

3.3.4 Mean streamwise vorticity

A series of contour plots of mean streamwise vorticity $\langle\omega_x\rangle$ for the same range of Reynolds numbers are shown in Figure 3.10, consisting of in total four noticeable vorticity cells, namely: (1) inner secondary vorticity cell, (2) outer secondary vorticity cell, and (3,4) a pair of solid-solid corner vorticity cells.

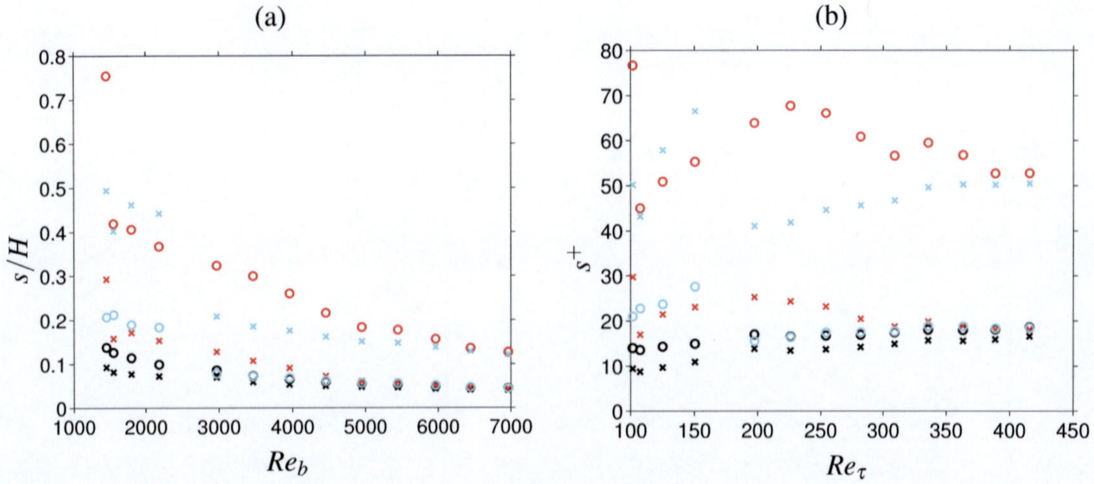

Figure 3.11: Positions of extrema of $\langle\omega_x\rangle(y,z)$: (a) external coordinates vs. Re_b and (b) wall coordinates vs. Re_τ. Colours: black, inner secondary vorticity cell; red, bottom-wall vorticity cell; cyan, side-wall vorticity cell. Symbols: ∘ distance to $z/H = -A$; × distance to $y/H = 1$ (black symbols) and $y/H = 0$ (red and cyan symbols).

Additionally, there are a number of mirror-vorticity layers on the no-slip walls that which we do not consider here. For convenience, the vorticity cell (3) which is directly above the side-wall is now referred as "side-wall vorticity cell", while the remaining vorticity cell (4) is now called "bottom-wall vorticity cell".

Similar to the bottom-wall secondary flow cell discussed earlier, the dimension and the position of the bottom-wall vorticity cell differ significantly for $Re_b < 1800$ and higher. This direct correspondence between the mean secondary flow and the streamwise vorticity patterns are reasonable for this marginal Reynolds number range, explained by the absence of scale-separation of motions.

As Reynolds number increases, those vorticity cells move towards the corresponding corners as shown in Pinelli et al. (2010) for the closed duct cases, except the outer secondary cell that is not in vicinity of any of the corners of the cross-section. The outcome of the outer secondary vorticity cell is unique: while the dimension and the peak location in y-direction are attracted towards the free-slip plane, the same quantities in z-direction remain qualitatively unchanged.

Once again, such trend is quantified in terms of the change in the peak locations as function of Reynolds number, shown in Figure 3.11 for the inner, side-wall/bottom-wall vorticity cells, while in Figure 3.12 for the outer vorticity cell.

It can be clearly seen that the outer secondary vorticity peak is located consistently at $|z/H| = 0.5$, while the distance from the free-slip plane is constantly around 14^+. Note that the peak distance from the free-slip plane coincides well with the radius of the typical buffer-layer vortex model suggested by Kim et al. (1987) ($\approx 15^+$), conjectured from their aforementioned rms profile of the streamwise vorticity fluctuation.

In order to investigate the source of those heterogeneous scaling properties from a viewpoint of the preferential locations of quasi-streamwise vortices, a vortex eduction study needs to be performed as in Uhlmann et al. (2007) and Pinelli et al. (2010). A brief summary of the applied vortex eduction technique based on the criteria proposed by Kida and Miura (1998) is provided here: for each cross-sectional plane of a sufficiently large number of instantaneous fields (typically

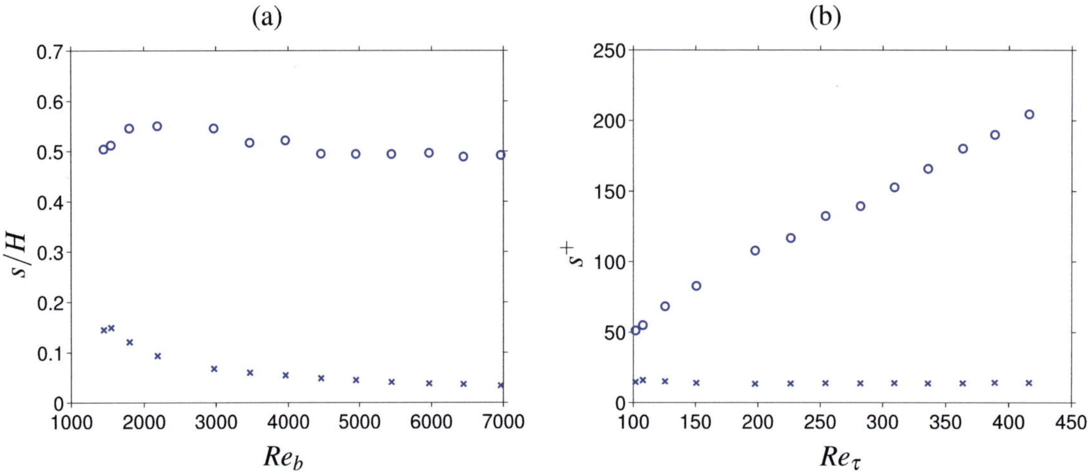

Figure 3.12: Positions of extrema of $\langle\omega_x\rangle(y,z)$ of outer secondary vorticity cell in: (a) external coordinates vs. Re_b and (b) wall coordinates vs. Re_τ. Symbols: $\circ$ distance to $z/H = -A$; $\times$ distance to $y/H = 1$.

> 1000), each local low pressure minima is identified as a candidate of possible vortex core; those candidates are then verified in terms of an additional swirl condition $D < 0$, where D is the discriminant of the cross-sectional (two-dimensional) velocity gradient tensor (i.e. $D = (\partial v/\partial y - \partial w/\partial z)^2/4 + (\partial v/\partial z)(\partial w/\partial y)$). Subsequently, The identified vortex centres are stored together with the corresponding local streamwise vorticity in order to categorise them according to the sign of ω_x at the centre location. Note that the educed vortex cores with weak vorticity levels (below 1% of the absolute maximum of all detected core intensities) are eliminated from the following statistics.

The results processed from 1000 instantaneous fields separated by $\approx 10\nu/u_\tau^2$ are shown in Figure 3.13(a-c): (a) probability of occurrence of positive vortex cores; (b) negative vortex core; and (c) the difference between positive and negative probabilities.

Firstly from the correspondence between Figure 3.13(c) and the mean streamwise vorticity profile (cf. Figure 3.13(e)), it can be said that the mean streamwise vorticity pattern is the statistical footprint of the most probable locations of the quasi-streamwise vortices, as of the closed duct counterparts. Secondly, the existence of tightly-concentrated preferable areas inside the mixed-boundary corners for the vortices rotating only in the direction of the inner secondary flow should be noted. This second finding is particularly remarkable, not only because of their much higher probability peak levels than any other part of the cross-sectional plane, but also due to the qualitative difference from the quiescent nature of the solid-solid corners. In other words, the quasi-streamwise vortices **avoid** those solid-solid corners, while they **prefer** the mixed-boundary corners only if they rotate in the direction of the mean inner secondary flow. The mixed-boundary corners also accommodate strong peaks of $\langle\omega_x\rangle_{\mathrm{rms}}$ (cf. Figure 3.13(d)), that is consistent with the persistent existence of quasi-streamwise vortices inside the corners detected by the current vortex eduction study.

Therefore, those mixed-boundary corner probability peaks and the mean streamwise vorticity fluctuation peaks, as well as the wall-scaling of the inner secondary flow cell peaks consistently imply that the inner secondary flow cell is a direct signature of quasi-streamwise vortices. Note that the

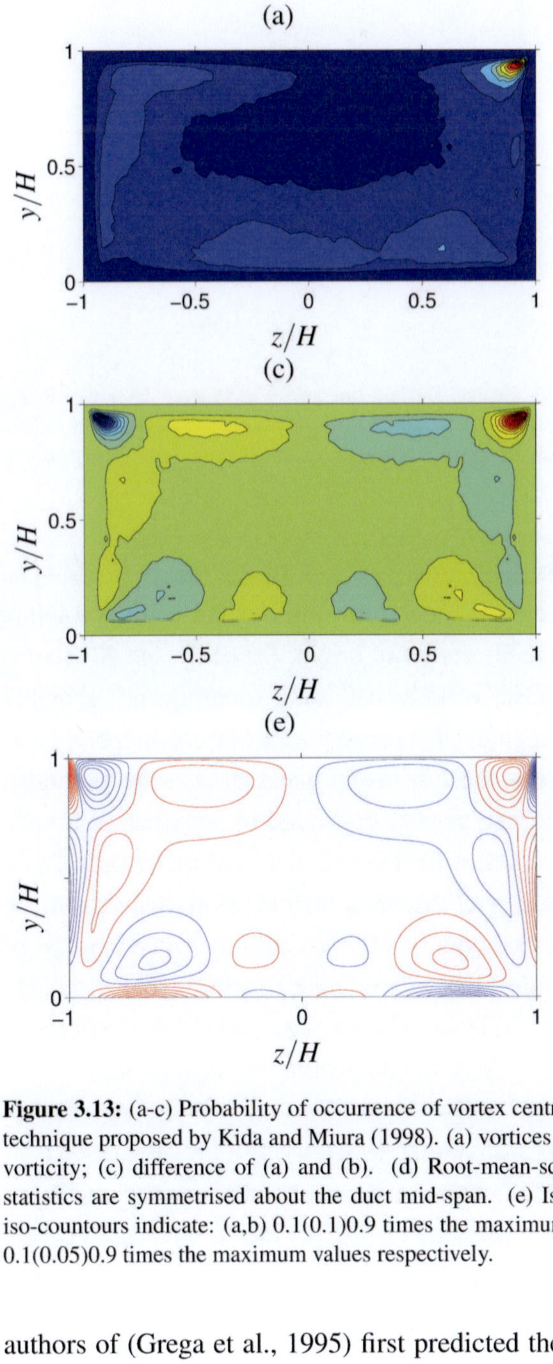

Figure 3.13: (a-c) Probability of occurrence of vortex centres for the open duct case with $Re_b = 2205$, detected by the technique proposed by Kida and Miura (1998). (a) vortices with *positive* streamwise vorticity; (b) *negative* streamwise vorticity; (c) difference of (a) and (b). (d) Root-mean-square of streamwise vorticity fluctuation $\sqrt{\langle \omega_x' \omega_x' \rangle}$. (c,d) statistics are symmetrised about the duct mid-span. (e) Iso-contours of $\langle \omega_x \rangle (y,z)$ (replotted for convenience). The iso-countours indicate: (a,b) 0.1(0.1)0.9 times the maximum values; (c) -0.9(0.1)0.9 times the maximum values; (d,e) 0.1(0.05)0.9 times the maximum values respectively.

authors of (Grega et al., 1995) first predicted the existence of the instantaneous inner secondary vortices inside the mixed-boundary corners, which is consistent with our findings in the current study, but also witnessed the difficulties to experimentally demonstrate the existence due to their weak intensities. In their follow-up studies in Hsu et al. (2000) and Grega et al. (2002), however, they modified their claim that the inner secondary flow only appears in the averaged velocity field, and therefore occurrence of instantaneous vortices is not expected.

Now let us turn our attention to the coherent structures responsible for the outer secondary vorticity cell shown in Figure 3.13(a,b).

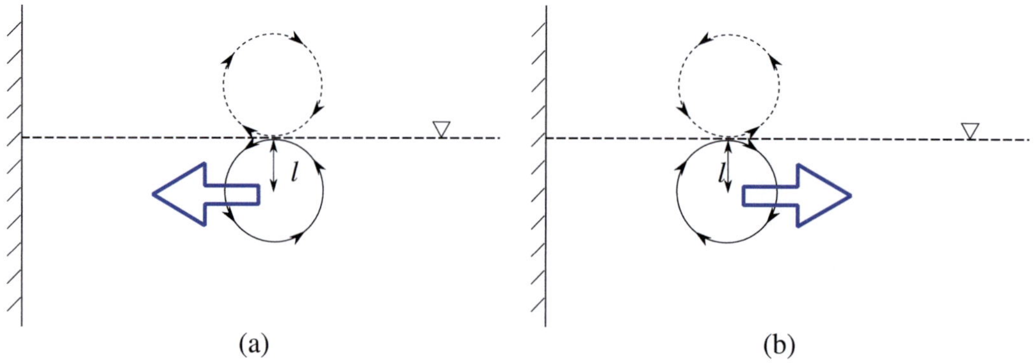

Figure 3.14: Conceptual sketch of mixed-boundary vortex dynamics. Dashed-line vortex represents the mirror image generated by free-slip boundary in order to satisfy impermeability constrain. Blue arrow indicates the induced motion due to the vortex dipole effect. Vortex with the sense of rotation corresponding to: (a) mean inner secondary vorticity cell; (b) outer secondary vorticity cell.

Firstly, it is interesting to observe that from the mixed-boundary corners towards the duct mid-span along the free-slip plane, there is a nose-shaped region of high population of the vortices rotating in the opposite direction to the ones accumulated in the corner of origin. Secondly, the distributed vortex core probability implies that the outer secondary vorticity cell is not a consequence of highly deformed large-scale vortices covering the whole region, but a consequence of a collection of near-wall vortices instead. Moreover, it is less likely that those free-slip plane vortices are generated on-site, due to the absence of constant high shear rate usually generated by presence of no-slip walls in vicinity, which is known as the key requirement for the generation of quasi-streamwise vortices (cf. Lee et al. (1990)). Therefore, those near-wall structures should be originated from the side and bottom-wall regions. Furthermore, since the bottom-wall becomes too distant in terms of wall units in case of sufficiently high Reynolds number, majority of those structures are expected to be originated from the side-wall regions eventually. This trend is already visible from the free-slip plane vorticity distribution that is extended smoothly from the ones from the side-walls.

Above observations all indicate a necessity of mechanisms of sorting and transporting vortices based on their sign of rotation in vicinity of the free-slip plane, which we will investigate next.

In order to simplify the problem, let us now consider a two-dimensional domain corresponding to the open duct cross-section, and placing a vortex inside one of the two mixed boundary corners (cf. Figure 3.14) representing a vortex generated on the side-wall boundary. Such simplification is justified by the fact that near-wall vortices are highly elongated in the streamwise direction in general.

We then refer to the numerical study of Orlandi (1990) that simulated the dynamics of two-dimensional vortex dipole interacting with no-slip and free-slip boundaries separately. One of the key findings of Orlandi (1990), which is useful for our current purpose, is that a vortex (dipole) located sufficiently close to a free-slip plane sees its mirrored image due to the imposed symmetric boundary condition. As a consequence, the vortex and its image forms a new dipole and is transported along the free-slip plane due the induced velocity from each other.

Consequently in our simplified 2D problem: if the vortex is rotation in the direction of inner secondary vortex, it will stay inside the corner and possibly be exposed to the complex non-linear cycle by interacting with the side-wall no-slip boundary; while if the rotational direction is as of the outer secondary vortex, then the vortex will move away from the corner along the free-slip plane and the movement is eventually constrained by the bulk geometry. Note that an evidence of such vortex-free-slip-plane interaction occurring in fully-developed three-dimensional turbulence can be seen from the numerical open channel study of Nagaosa (1999) (their Figure 6 and 7).

Here we attempt to approximate a range of influence of the free-slip plane to nearby vortices by supposing the break-even point to be where the characteristic velocity scale of the near-wall turbulent motions matches with the dipole-induced velocity.

First, let us approximate the characteristic turbulent velocity scale from $\langle\omega_x\rangle_{\text{rms}}$ with some typical length-scale, such as mean radius of the vortical motion l:

$$V_{\text{turb}} \approx \langle\omega_x\rangle_{\text{rms}} l \,. \tag{3.4}$$

Second, the induced velocity scale from the mirror image can be written as:

$$V_{\text{ind}} = \frac{\Gamma}{2\pi d} \,, \tag{3.5}$$

where Γ is circulation of the vortex, while d is the distance between the vortex and its image. Let us assume

$$\Gamma = \langle\omega_x\rangle_{\text{rms}} l^2 \pi \,, \tag{3.6}$$

then the induced velocity can be approximated as:

$$V_{\text{ind}} \approx \frac{\langle\omega_x\rangle_{\text{rms}} l^2}{2d} \,. \tag{3.7}$$

Recall the break-even point set at where $V_{\text{turb}} \approx V_{\text{ind}}$, therefore

$$\langle\omega_x\rangle_{\text{rms}} l \approx \frac{\langle\omega_x\rangle_{\text{rms}} l^2}{2d} \,, \tag{3.8}$$

then finally we obtain

$$d \sim l \,. \tag{3.9}$$

This attempted order-of-magnitude analysis implies that the influence range should be in the order of vortex radius, that is approximately 15 wall units in wall-bounded flow as mentioned earlier. Therefore in a fully-developed open duct turbulence, only those vortices that are inside a very thin layer immediately below the free-slip plane with the height in few tens of wall units should be sorted and transported according to their rotational directions. Consequently, the mixed-boundary corners are highly populated by the vortices with a certain sense of rotation that are trapped inside the corner region, whereas other vortices in the opposite rotation move away along the free-slip plane. As a result, the distribution of the vortices corresponding to the outer vorticity cell is expected to have a dimension in the inner length-scale in y-direction (i.e. $\approx$diameter of the vortices), whereas in the outer length-scale in z-direction, due to the limitation upon the transport distance imposed by the duct geometry. Note that the predicted length-scales of the distribution are consistent with our earlier findings.

3.3.5 Other important scaling

3.3.5.1 Global characteristics

Secondary flow intensity is another important quantity for both scientific and engineering interests, but any concrete consensus is yet to be formed like many other aspects of duct turbulence.
For the square closed duct turbulence, direct numerical simulations of Gavrilakis (1992) set an important reference value that the maximum secondary flow intensity is approximately at 1.9% of the bulk velocity, while direct numerical simulations of Raiesi et al. (2011) reported the same quantity of 4%.
For open duct turbulence, Tominaga et al. (1989) reported that the maximum intensity of the secondary velocity in their open duct experiments is 1.5% of the maximum streamwise velocity. Also from experiments, Hsu et al. (2000) reported the intensity of the inner secondary flow is in the order of 1% of the bulk velocity. On the other hand, Joung and Choi (2010) determined the maximum intensity of the inner secondary flow numerically, that is about 3% of the bulk velocity. Figure 3.15 shows both cross-plane averaged and maximum intensities of the mean secondary flow defined respectively by

$$U_\perp = \left[\frac{1}{2AH^2}\int_{-AH}^{+AH}\int_0^H \left(\langle v\rangle^2 + \langle w\rangle^2\right) dy\, dz\right]^{1/2}, \tag{3.10}$$

and $\max\{(\langle v\rangle^2 + \langle w\rangle^2)^{1/2}\}$ from the current study. Note that the square closed duct results from Pinelli et al. (2010) ($Re_b \leq 3500$) and Zhang et al. (2015) ($Re_b \approx [2205, 4840, 7630, 10710]$) are also included, alongside our new closed duct simulation results of $Re_b \approx [4000, 4500, 5000, 5500, 6000, 6500, 7000]$.
First of all, it can be seen that the cross-sectional average of the open duct secondary flow intensity $U_\perp$ scales fairly well with the bulk velocity u_b, whereas in case of the closed duct flow with $Re_b > 4000$, it does not seem to scale with neither bulk nor wall velocities. It turns out that $U_\perp$ can be made roughly constant over the range of Reynolds numbers from approximately 4000 up to the largest currently available values in DNS by normalising with the velocity scale $\sqrt{u_\tau^3/u_v}$ (cf. Figure 3.15(c)).
The maximum closed duct secondary flow intensity scales very well with u_b, which is around 2% after normalisation that is consistent with the value reported by Gavrilakis (1992). On the contrary, the maximum open duct secondary flow intensity increases monotonically and reaches as high as 7% of u_b at $Re_b = 7000$, where the maximum location is always found on the free-slip plane. The observed ratio to the bulk velocity is significantly higher than the values obtained from experimental measurements, and this discrepancy is believed to be a direct consequence of the difficulty involving precise measurements directly on the free surface.
Scaling properties of cross-plane averaged velocity fluctuations are also interesting quantities to look at. Figure 3.16 show the integral measure of the cross-streamwise ($\langle v'v'\rangle$ and $\langle w'w'\rangle$) as well as the streamwise ($\langle u'u'\rangle$) velocity fluctuations respectively as function of Reynolds number. Moreover, the integral measure of the correlation of the two components of the cross-streamwise velocity fluctuation ($\langle v'w'\rangle$) is presented in Figure 3.17.

It can be seen the cross-streamwise velocity fluctuation intensities:

$$U_{\mathrm{rms}\perp} = \left[\frac{1}{2AH^2}\int_{-AH}^{+AH}\int_{0}^{H}\langle v'^2\rangle + \langle w'^2\rangle dydz\right]^{1/2} \tag{3.11}$$

of both open and closed duct flow scale very well with u_b, whereas the corresponding streamwise fluctuation intensities:

$$U_{\mathrm{rms}} = \left[\frac{1}{2AH^2}\int_{-AH}^{+AH}\int_{0}^{H}\langle u'^2\rangle dydz\right]^{1/2} \tag{3.12}$$

scales again neither u_b nor u_τ, but it turns out that the quantity scales with a combination of all three aforementioned velocity scales: $(\sqrt{u_b u_\nu} + u_\tau)$. Furthermore, the integrated cross-streamwise velocity fluctuation correlation:

$$U_{\sqrt{\langle v'w'\rangle}} = \left[\frac{1}{2AH^2}\int_{-AH}^{+AH}\int_{0}^{H}|\langle v'w'\rangle| dydz\right]^{1/2} \tag{3.13}$$

scales fairly well with the friction velocity u_τ. Note that there is a noticeable discrepancy between ours and the data deduced from Zhang et al. (2015) for U_{rms}, whereas those data match satisfactory for the cross-streamwise quantities. Further investigation is therefore required to determine the cause of the discrepancy.

Finally, Reynolds number scaling of cross-section averaged enstrophy:

$$\Omega = \frac{1}{2AH^2}\int_{-AH}^{+AH}\int_{0}^{H}(\langle\omega_x\rangle^2/2)\, dy\, dz \tag{3.14}$$

is also shown in Figure 3.18, again with no sign of straightforward scaling both by u_b and u_τ, but it turns out that the closed duct values scale with again a combination of the three velocity scales $u_b u_\tau^2/(u_\nu H^2)$ for $Re_b > 2000$.

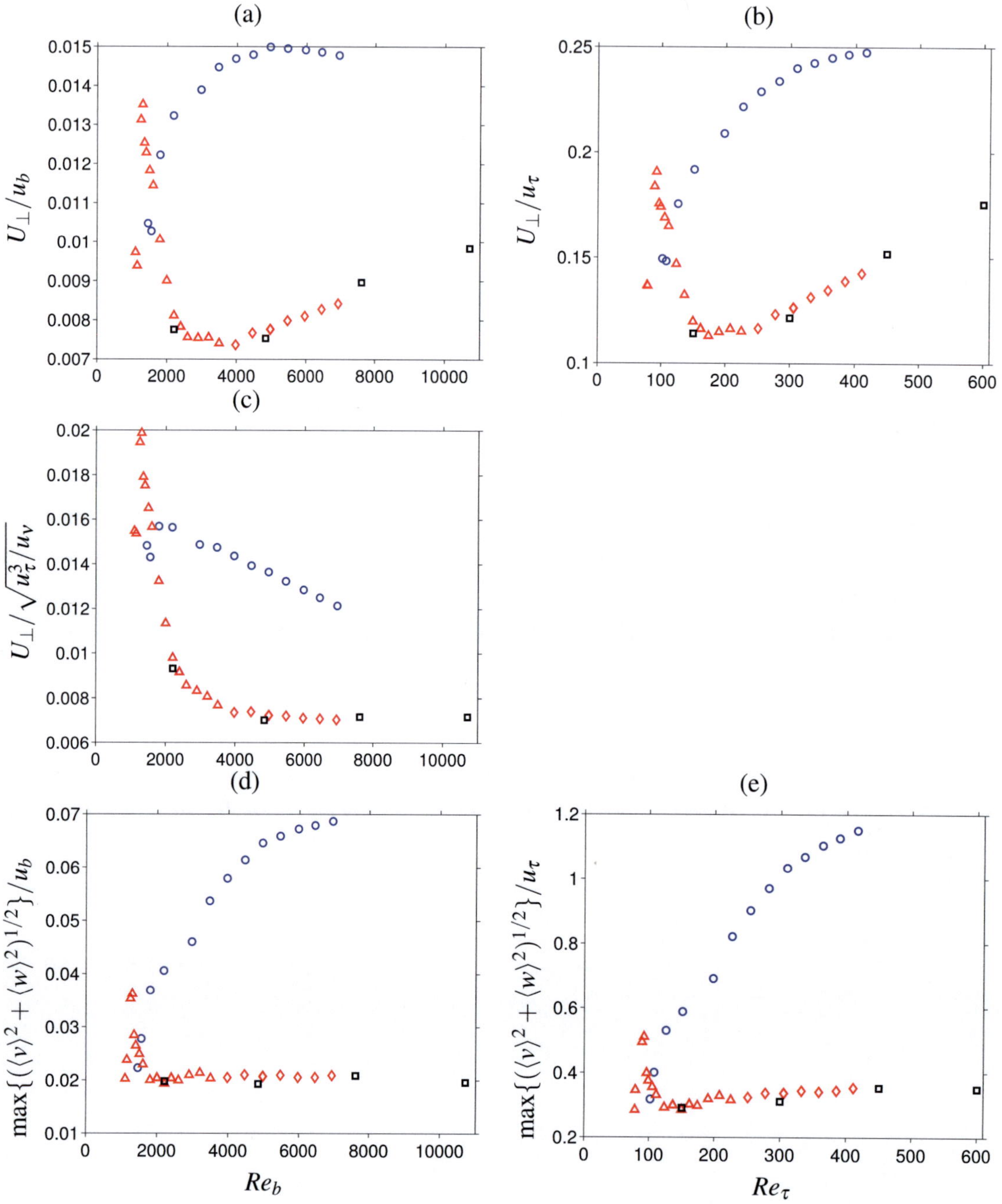

Figure 3.15: (a-c) Intensities of cross-section averaged mean secondary velocity $U_\perp = [\frac{1}{2AH^2}\int_{-AH}^{+AH}\int_0^H (\langle v\rangle^2 + \langle w\rangle^2)\, dydz]^{1/2}$. (d,e) Intensities of maximum mean secondary velocity $\max\{(\langle v\rangle^2 + \langle w\rangle^2)^{1/2}\}$. The values are normalised by: (a,d) u_b and vs. Re_b; (b,e) u_τ and vs. Re_τ; (c) $\sqrt{u_\tau^3/u_\nu}$ and vs. Re_b. Symbols indicate: $\circ$, open duct of current study; $\triangle$, closed duct of Pinelli et al. (2010); $\diamond$, closed duct of current study; $\square$, closed duct of Zhang et al. (2015).

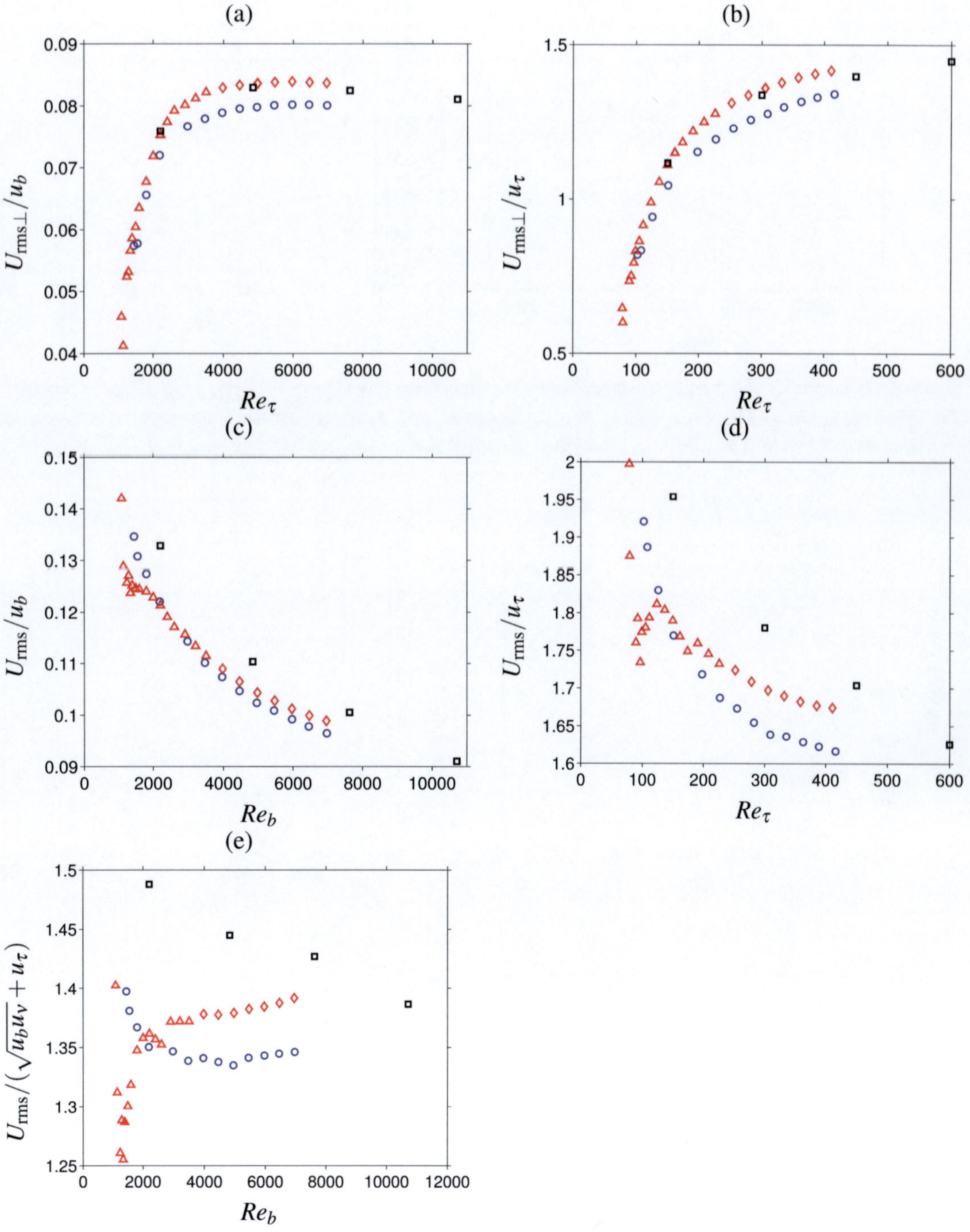

Figure 3.16: Intensities of cross-plane averaged: (a,b) cross-streamwise rms velocity $U_{\mathrm{rms}\perp} = [\frac{1}{2AH^2}\int_{-AH}^{+AH}\int_0^H \langle v'^2\rangle + \langle w'^2\rangle dydz]^{1/2}$; (c-e) streamwise rms velocity $U_{\mathrm{rms}} = [\frac{1}{2AH^2}\int_{-AH}^{+AH}\int_0^H \langle u'^2\rangle dydz]^{1/2}$. The values are normalised by: (a,c,) u_b and vs. Re_b; (b,d) u_τ and vs. Re_τ; (e) $\sqrt{u_b u_\nu}+u_\tau$ vs. Re_b. Symbols indicate: $\circ$, open duct of current study; $\triangle$, closed duct of Pinelli et al. (2010); $\diamond$, closed duct of current study; $\square$, closed duct of Zhang et al. (2015).

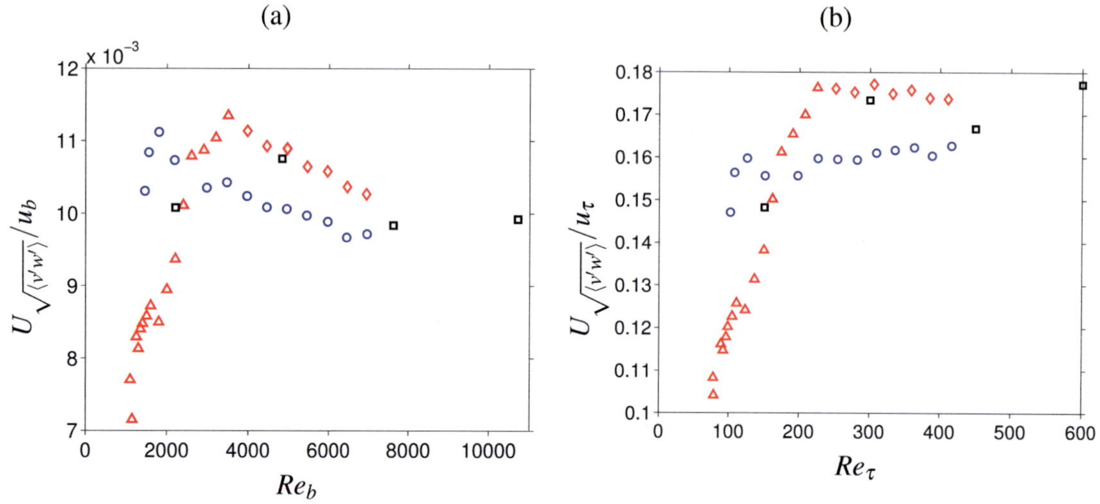

Figure 3.17: Intensities of cross-plane averaged cross-correlation velocity $U_{\sqrt{\langle v'w'\rangle}} = [\frac{1}{2AH^2}\int_{-AH}^{+AH}\int_0^H |\langle v'w'\rangle| dydz]^{1/2}$. The values are normalised by: (a) u_b and vs. Re_b; (b) u_τ and vs. Re_τ. Symbols indicate: ○, open duct of current study; △, closed duct of Pinelli et al. (2010); ◇, closed duct of current study; □, closed duct of Zhang et al. (2015).

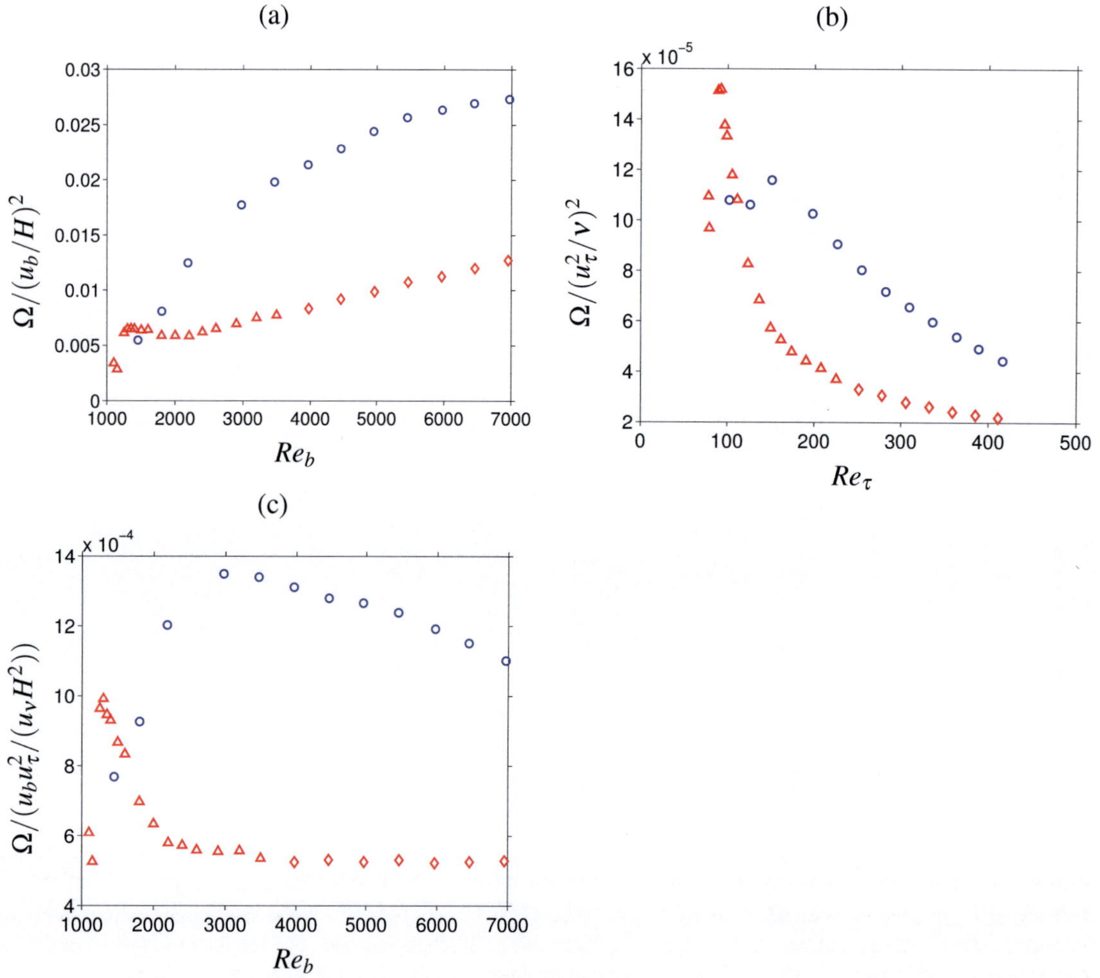

Figure 3.18: Cross-section averaged Enstrophy $\Omega = \frac{1}{2AH^2}\int_{-AH}^{+AH}\int_0^H (\langle\omega_x\rangle^2/2)\ dy\ dz$ normalised by (a) bulk units $(u_b/H)^2$; (b) wall units $(u_\tau^2/\nu)^2$; (c) mixed units $(u_b u_\tau^2/(u_\nu H^2))$. Symbols indicate: ○, open duct of current study; △, closed duct of Pinelli et al. (2010); ◇, closed duct of current study.

3.3.5.2 Dip-phenomenon

Despite our long-time recognition of the dip-phenomenon — occurrence of the maximum streamwise velocity somewhat below free surface— that spans well over one hundred years (cf. Stearns (1883)), its detailed mechanism as well as the scaling properties are still open questions to be answered. Consequently, structural interpretations of such phenomena are rare in the literature, though fortunately few do exist such as the one suggested by Nezu and Nakagawa (1993). They conjecture the phenomenon, unfortunately without any strong evidence, as a consequence of low-momentum fluid transported from the side-walls of an open duct along the free surface by secondary flow, relocating the high-momentum fluid around the region towards duct core. As a result, it submerges the maximum velocity location below the free surface for a certain distance, and if the duct aspect ratio is below a critical value, the phenomena can occur even at the duct mid-span. A phenomenological evidence of such large-scale migration events of the side-wall low-momentum fluid across the free surface was later presented by means of an instantaneous visualisation in Grega et al. (1995) (cf. their figure 8).
Experimental studies of Tominaga et al. (1989) reported that the submerged distance is about $0.2-0.3H$ from the free surface at the duct mid-span, if the aspect ratio is below $A_{crit} \approx 2.5$.
More recently, Wang et al. (2001) proposed an empirical formula that estimates the impact of the dip-phenomenon as a function of distance from the closest side-wall $z'' = AH - |z|$, viz.

$$\frac{y_{\max}}{H} = 0.44 + 0.212\frac{z''}{H} + 0.05\sin\left(\frac{2\pi z''}{2.6H}\right), \tag{3.15}$$

where $y_{\max}$ is the location of the maximum streamwise velocity. The critical aspect ratio predicted by their formula is approximately at 2.63, and the submerged distance for our current aspect ratio ($A = 1$) is around $0.31H$.
An alternative method of estimation was proposed by Yang et al. (2004) based on their finding that the streamwise deviation from the well-know log law is linearly proportional to the logarithmic distance from the free surface. Their model states:

$$\frac{y_{\max}}{H} = \frac{1}{1+\alpha}, \tag{3.16}$$

with $\alpha = 1.3\,\exp(-z''/H)$ where α is a dip-correction factor. Consequently, the predicted submerged distance when $A = 1$ is approximately at $0.32H$. Note that although its fair accuracy and great ease of use, their model has one major drawback: incapability to predict the critical aspect ratio (i.e. $y_{\max}$ never strictly reaches to zero).
Figure 3.19 shows the submerged distance d as a function of Reynolds number from our current dataset. It can be seen that the dip-phenomenon does not initially occur at $Re_b = 1450$, then appears rather suddenly for $Re_b \geq 1550$ with the distance slightly shallower than the predicted values by the aforementioned models, then gradually converges towards the terminal value of $\approx 0.33H$. Note that the converged distance is strictly scaling with bulk units, indicating a predominant influence from the outer-scaling large-scale motions.
In order to verify the structural conjuncture of the dip-phenomenon by Nezu and Nakagawa (1993), it is necessary to capture the preferential locations of high-/low-velocity streaks over the duct

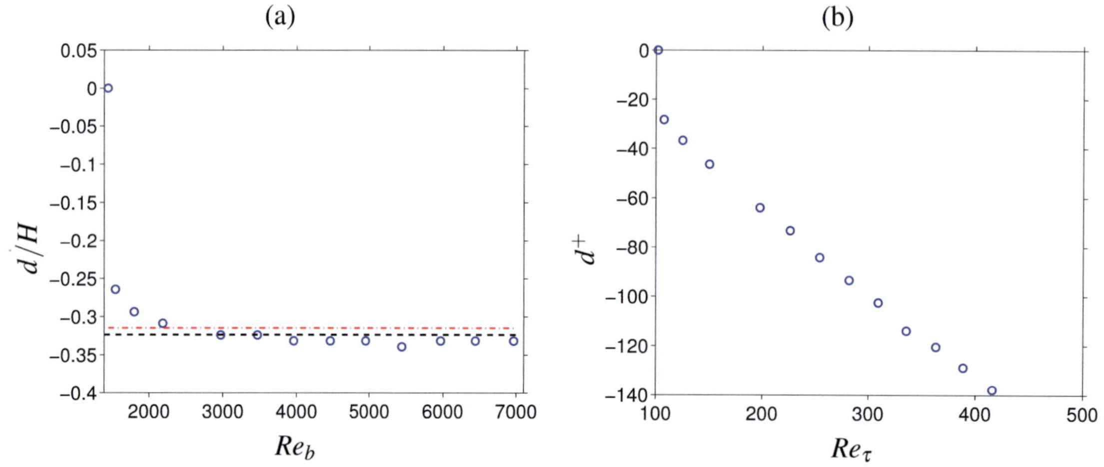

Figure 3.19: (a,b) Submerged distances d from the free-slip plane $y/H = 1$ to the location of the maximum mean streamwise velocity $\langle u \rangle$ along the duct mid-span $z/H = 0$, respectively in: (a) external coordinates vs. Re_b and (b) wall coordinates vs. Re_τ. Red dotted dash and black dashed lines indicate the predicted values by the models of Wang et al. (2001) and Yang et al. (2004) respectively.

cross-section. Furthermore, it is also desirable to show such probabilistic distributions presented separately in terms of their length-scale to demonstrate the aforementioned scaling properties as a consequence. For that purpose, we employ a coherent structure eduction technique considering an instantaneous distribution of streamwise velocity fluctuation u' on each two-dimensional duct cross-section. Note that the eduction technique that is considered here was first introduced and employed by Nakatsuji (2012) (MSc. thesis, Univ. Osaka), however the information was unfortunately only available in Japanese therefore will be detailed here.

We educe a set of enclosed cross-sectional areas defined by a given threshold value u_{th}, in order to characterise the locations and sizes of individual high-/low-velocity streaks at each instance. Full definition of the eduction criteria is as follows:

$$u'/\sigma_u > u_{\mathrm{th}} \tag{3.17}$$

and

$$u'/\sigma_u < -u_{\mathrm{th}} \tag{3.18}$$

for high-/low-velocity streaks respectively. Note that the employed normalisation factor σ_u is rms value of u' based on its volume-averaged quantity (represented as $\langle \phi \rangle_V$) per instantaneous field, viz.

$$\sigma_u = \sqrt{\langle u'^2 \rangle_V}\ . \tag{3.19}$$

Representative locations of those detected low-/high-velocity streaks are defined as:

$$y_{\mathrm{c}} = \frac{\int_{\mathscr{A}_{\mathrm{s}}} y u' \mathrm{d}S}{\int_{\mathscr{A}_{\mathrm{s}}} u' \mathrm{d}S},\ z_{\mathrm{c}} = \frac{\int_{\mathscr{A}_{\mathrm{s}}} z u' \mathrm{d}S}{\int_{\mathscr{A}_{\mathrm{s}}} u' \mathrm{d}S}\ . \tag{3.20}$$

Consequently, a characteristic length-scale of each educed structure can be computed from their cross-sectional area $\mathscr{A}_{\mathrm{s}}$, viz.

$$l_{\mathscr{A}} = \sqrt{\mathscr{A}_s} \,. \tag{3.21}$$

In order to examine the effect of the threshold value choice, four different values were applied ($u_{th} = [0.25, 0.5, 1, 2]$) on 1000 instantaneous fields separated by $\approx 10\nu/u_\tau^2$, that were sampled from each four representative Reynolds number $Re_b = [1450, 2205, 3000, 5000]$. Consequently, the threshold value as well as Reynolds number dependency of: (a,b) average maximum length-scale $\langle l_{\mathscr{A}\max}\rangle$; (c,d) average length-scale $\langle l_{\mathscr{A}}\rangle$; (e,f) average number of the educed structures $\langle N_s\rangle$ are summarised in Figure 3.20.

Firstly, it can be seen that the values of both $\langle l_{\mathscr{A}\max}\rangle$ and $\langle l_{\mathscr{A}}\rangle$ decrease monotonically as u_{th} increases, caused by the fact that a stricter threshold value results smaller areas. Secondly, the maximum length-scale $\langle l_{A\max}\rangle$ is very much Reynolds number independent, which indicates that those detected large structures correspond to the large-scale bulk-scaling structures mentioned previously. Thirdly, the averaged number of detected structures first increases then decreases, indicating a trend that: at first structures are artificially connected when the threshold value is small; then correct skeletons of individual structures are started to be revealed with increasing threshold value; but eventually small structures starts to be dismissed if the threshold value is too high. It is therefore reasonable to choose a value corresponding to the number of the detected structures is near its peak: our choice here is at $u_{th} = 0.5$.

Those detected structures are then categorised into four groups in terms of their $l_{\mathscr{A}}$, namely: (i) $l^+_{\mathscr{A}} < 25$; (ii) $25 < l^+_{\mathscr{A}} < 50$; (iii) $l^+_{\mathscr{A}} > 50$, $l_{\mathscr{A}}/H < 0.5$; (iv) $l_{\mathscr{A}}/H > 0.5$, and the probability density function of their locations are computed separately. The outcomes for $Re_b = 1450$ are presented in Figure 3.21, while Figure 3.22 shows for $Re_b = 2205$ (note: the corresponding results for $Re_b = [3000, 5000]$ are in Appendix B).

First, let us consider the preferential locations of the smallest length-scale (i). It can be seen that those near-wall streaks with the inner-scaling length-scales are concentrated inside the mixed-boundary corners, drawing a consistent picture with the inner secondary flow with the inner-scaling dimensions as discussed earlier. On the contrary, the largest streaks with the outer-scaling dimension (iv) are found to be preferentially located in the regions where the mean secondary flow vortices —not to be confused with the instantaneous near-wall vortices— transport high- and low-momentum fluids towards and away from the solid-wall boundaries respectively. Distributions of the two largest length-scales (iii) and (iv) are qualitatively similar for $Re_b = 1450$, whereas their distributions significantly differ for $Re_b = 2205$, implying the development of some degree of scale separation of motions taking place for this range of Reynolds number already. It is also worth noting that the peak locations above the bottom-wall boundary of the length-scale (iii) for $Re_b = 2205$ capture the aforementioned five-streak state of the flow quite well.

Most important of all for the current purpose, a special attention should be placed upon the low-velocity streak distribution near the free-slip plane and duct mid-span ($y/H \approx 0.9$, $z/H \approx 0$). For $Re_b = 1450$ case, low-velocity streaks are not preferentially located around the concerned region (cf. Figure 3.21(b,d,f,h)), whereas for $Re_b = 2205$ (and higher), band-like signatures of existence parallel to the free-slip plane and originated from the side-wall regions can be observed (cf. Figure 3.22(d,f)). Furthermore, the length-scale of the structures that are predominantly found in the region of concern coincides well with the scale of the dip-phenomenon $d/H \approx 0.33$ that was discussed earlier. Please recall the length-scale is referred to as (iii) and defined by $l^+_{\mathscr{A}} > 50$, $l_{\mathscr{A}}/H < 0.5 \sim 0.33$.

In summary, this study verifies the structural conjecture suggested in Nezu and Nakagawa (1993) that the dip-phenomenon is a consequence of the low-velocity streaks transported from the duct side-walls, by demonstrating that low-velocity streaks are indeed preferentially located in the duct mid-span region near the free-slip plane only when the dip-phenomenon is taking place. Moreover, a strong link between the scaling of the dip-phenomenon and the outer-scaling of those low-velocity streaks preferentially located in the region was demonstrated.

3.4 Conclusion

Turbulent secondary motions in a straight open duct with the rectangular cross-section with the aspect ratio at unity was studied in this chapter, by means of direct numerical simulations up to the bulk Reynolds number at 7000. Consequently, our key findings can be summarised as follows.

First, the lowest sustainable Reynolds number of open duct flow with this aspect ratio was determined to be at $Re_b = 1450$ ($Re_\tau \approx 102$), where no temporal alternation of secondary flow orientation was observed in contrast to the corresponding closed duct marginal turbulence. Also, the observed minimum Reynolds number limit was significantly higher than the one determined previously at $Re_b = 1077$ ($Re_\tau \approx 77$), and the following examination of the mean flow field illustrated that the existence of additional mixed-boundary corner structures are responsible for the premature relaminarisation.

Second, the unique mean vorticity patterns were found to be the statistical footprint of the most probable locations of the quasi-streamwise vortices, therefore their dimensions shrink and peak locations move towards nearby no-slip walls as Reynolds number increases, whereas the mean flow pattern remains Reynolds-independent as for the corresponding closed duct flow. However, such trend was not applicable for the wall-scaling mean inner secondary flow cell peak locations and radius, as well as for the Reynolds-independent spanwise dimension of the outer secondary vorticity cells. We performed a coherent structure analysis to investigate the heterogeneous scaling properties, and revealed that the wall-scaling of the inner secondary cell radius is due to its direct correspondence with near-wall quasi-streamwise vortices of a certain sense of rotation, whereas the bulk-scaling of the outer secondary vorticity cell was found to be a consequence of the opposite-signed vortices spread broadly along the free-slip plane. Those observations, together with the non-existence of constant high shear near the free-slip plane necessary for vortex generation, consistently indicated an existence of some kind of sorting/transporting mechanism of vortices, that is based on the vortices' rotational direction and is only active in the vicinity of the free-slip plane. A possible explanation was then conjectured essentially as an inviscid two-dimensional process, that is based upon the previous numerical study on the vortex dipole dynamics influenced by the different boundary conditions.

Third, scaling of various cross-sectional averaged quantities were also discussed, and it was found that the global secondary flow intensity $U_\perp$ of open duct flow scales with u_b, in contrast to the integral measure of the closed duct configuration for $Re_b > 4000$ where it scales with $\sqrt{u_\tau^3/u_\nu}$. Note additionally that based on the observation $Re_\tau/Re_b \approx c$ where c is a constant, we can argue that $u_\tau \approx c \cdot u_b$ for the considered range of Reynolds number, therefore $\sqrt{u_\tau^3/u_\nu} \approx u_b\sqrt{u_b/u_\nu}$. Similarly, global scaling properties of the cross-streamwise, as well as the streamwise velocity fluctuation were also examined, and it was found that the former scales excellently with u_b in both open and closed duct flow, whereas the latter was found to scale with $(\sqrt{u_b u_\nu} + u_\tau)$. Finally, the integral measure of streamwise enstrophy was investigated, and it was found to scale with again a mixed velocity scale $u_b u_\tau^2/(u_\nu H^2)$ for the closed duct.

Finally, outer-scaling of velocity dip-phenomenon was confirmed by fully-resolved direct numerical simulations for the first time, and the previous hypothesis of the phenomenon to be a consequence of large-scale side-wall low-velocity streak ejection events was confirmed by a coherent structure eduction technique which takes into account the individual size of the educed coherent structures. It was found that the occurrence of mean dip-phenomenon can be explained by the existence of large-scale low-velocity streaks in the duct mid-span region near the free surface, and the corresponding submerged distance was found to be linked to the cross-sectional dimension of those streaks.

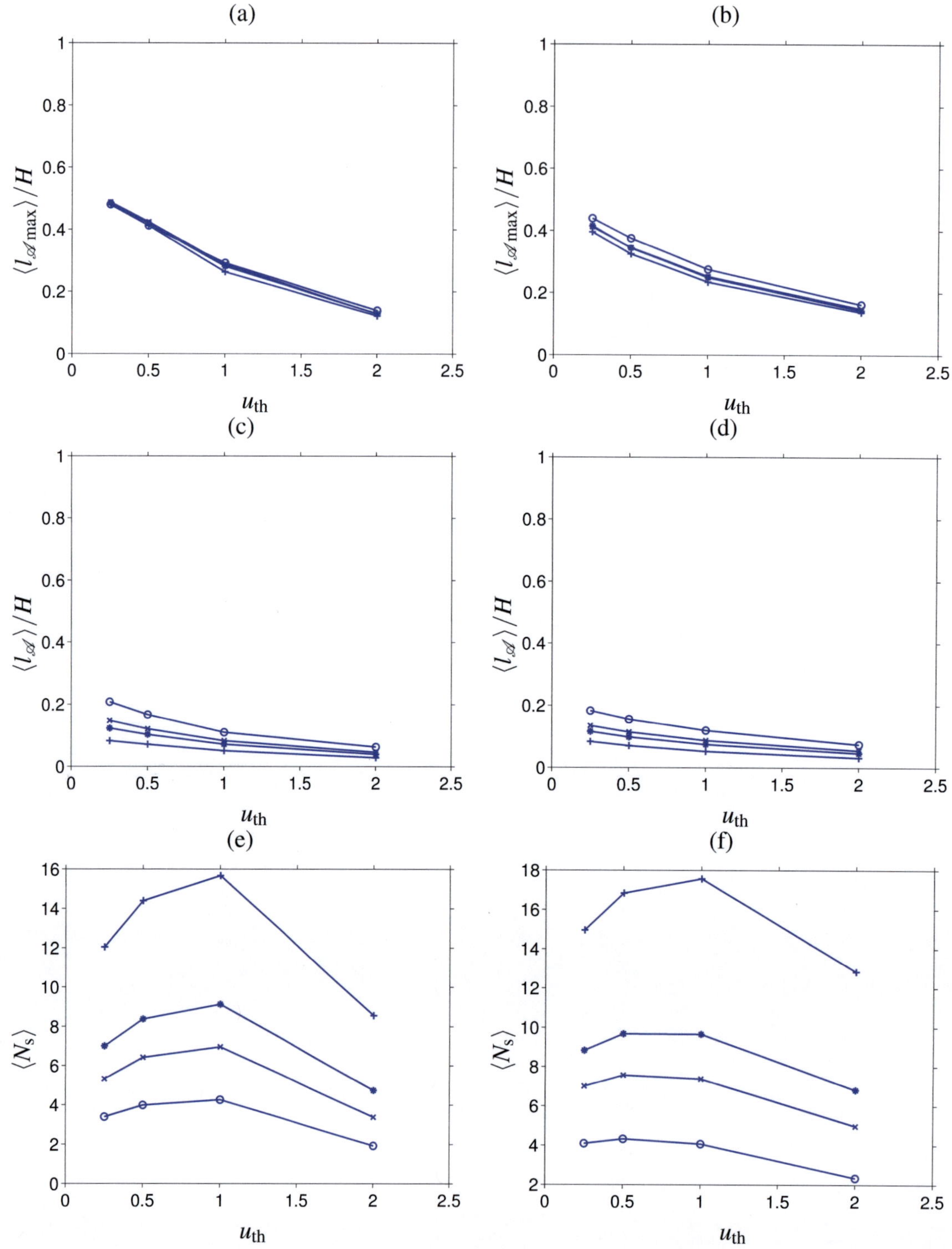

Figure 3.20: Ensemble average of: (a,b) maximum length-scale; (c,d) length-scale; (e,f) number of velocity streaks per cross-section as a function of threshold velocity u_{th}. (a,c,e) high-velocity streaks; (b,d,f) low-velocity streaks. Symbols indicate different Reynolds numbers: $\circ$, $Re_b = 1450$; $\times$, $Re_b = 2205$; $*$, $Re_b = 3000$; $+$, $Re_b = 5000$.

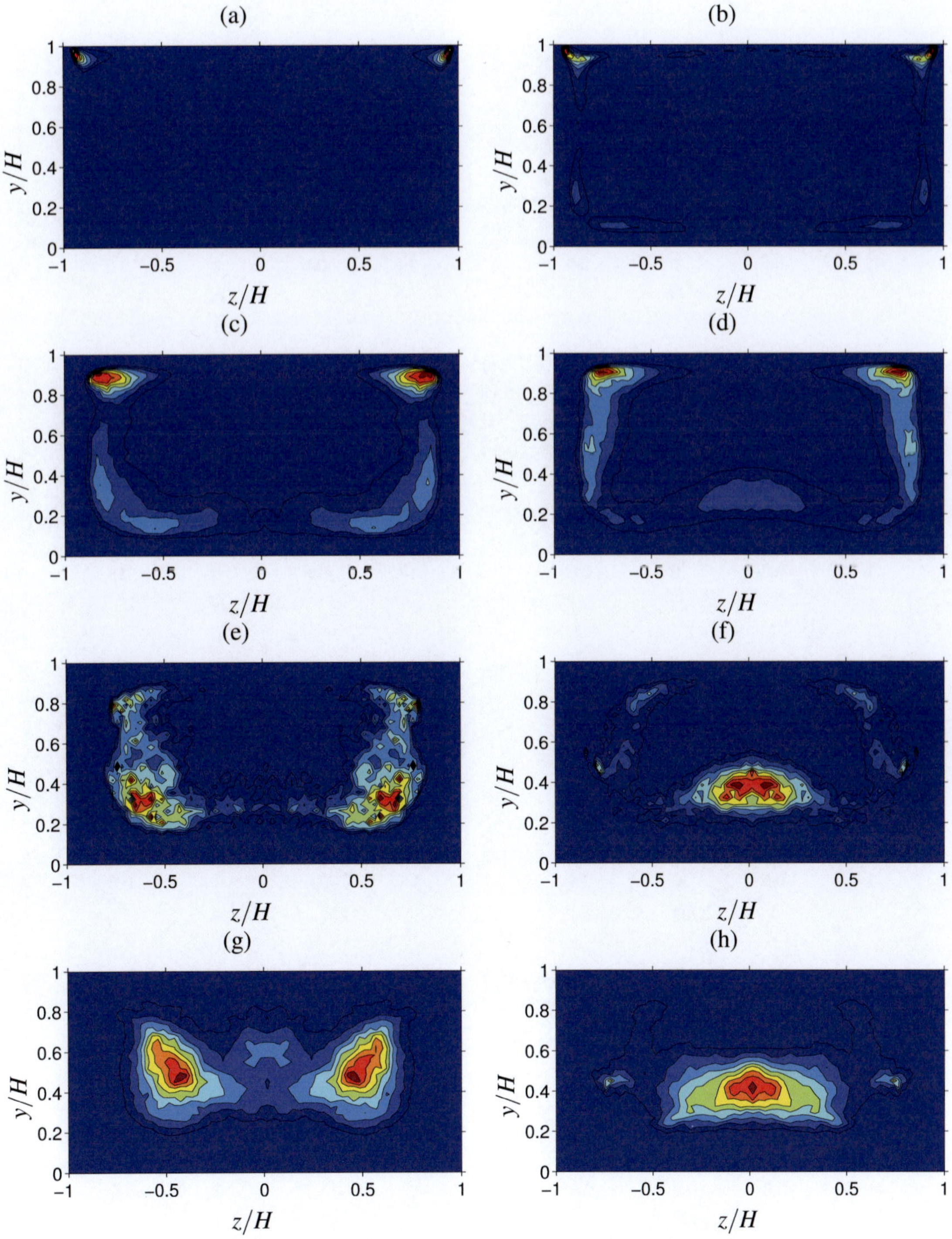

Figure 3.21: Probability density function of the location of: (a,c,e,g) high-; (b,d,f,h) low-velocity streaks. Velocity streaks are classified by their length-scale l_A as follows: (a,b) $l^+_{\mathscr{A}} < 25$; (c,d) $25 < l^+_{\mathscr{A}} < 50$; (e,f) $l^+_{\mathscr{A}} > 50$, $l_{\mathscr{A}}/H < 0.5$; (g,h) $l_{\mathscr{A}}/H > 0.5$. Contour lines indicate 0.1(0.1)0.9 times the maximum values, and statistics are symmetrised about the duct mid-span. Reynolds number is at $Re_b = 1450$, and the employed threshold value is at $u_{\text{th}} = 0.5$.

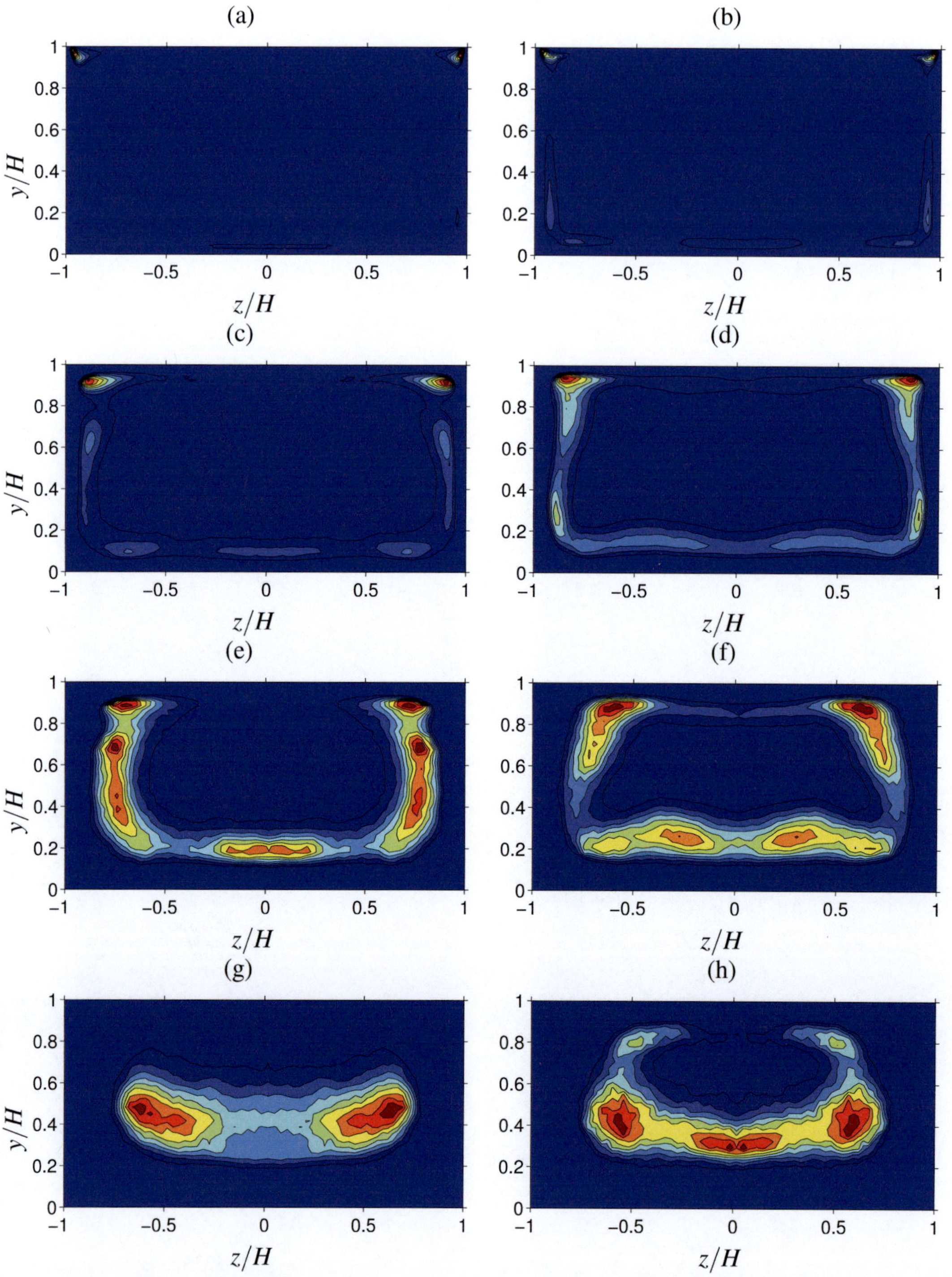

Figure 3.22: Probability density function of the location of: (a,c,e,g) high-; (b,d,f,h) low-velocity streaks. Velocity streaks are classified by their length-scale l_A as follows: (a,b) $l_A^+ < 25$; (c,d) $25 < l_A^+ < 50$; (e,f) $l_A^+ > 50$, $l_A/H < 0.5$; (g,h) $l_A/H > 0.5$. Contour lines indicate 0.1(0.1)0.9 times the maximum values, and statistics are symmetrised about the duct mid-span. Reynolds number is at $Re_b = 2205$, and the employed threshold value is at $u_{\mathrm{th}} = 0.5$.

Chapter 4

Aspect ratio dependence in open duct secondary flow

4.1 Introduction

The main objective of this chapter is to investigate the influence of the duct aspect ratio in the open duct secondary flows. Analogous to the previous chapter on the Reynolds number dependence, the aspect ratio dependence will be investigated with a special emphasis upon the turbulent coherent structure analysis.

First, the minimal sustainable Reynolds number limits of the open duct turbulence with two different aspect ratios $A = [0.5, 2]$, which are twice as narrower or wider as our reference marginal open duct case with $A = 1$, will be investigated. Detailed comparisons will be provided with the reference marginal open duct case, and our main interest in this part of the current investigation is to examine whether the minimal number of the coherent structures as well as their arrangement in the marginal open duct turbulence are influenced by the modified aspect ratio.

Subsequently, the flow Reynolds number will be increased to, and fixed at $Re_b = 2205$ where the turbulence is fully developed, whereas the aspect ratio A will be varied between 0.5 and 8. Accompanying closed duct simulations with the same Reynolds number and the aspect ratio ranging from 2 to 8 were also performed for the purpose of this study. Their statistics and detailed comparisons with the open duct data will be presented. Our main interest in this part of the investigation is to verify quantitatively the earlier finding of some experimental investigations (e.g. Tominaga et al. (1989)) that the mean secondary flow structures stay in the regions relatively close to the duct side-walls, even with an aspect ratio which is significantly larger than unity.

Lastly, consequence of the streamwise velocity dip-phenomenon in the wider aspect ratio open duct flow will be analysed, together with its Reynolds number dependence in the open ducts with $A = [0.5, 2]$.

A	Re_b	Re_τ	$M_x \times M_y \times M_z$	Δx^+	$\max\{\Delta y^+, \Delta z^+\}$	$t_{\text{stat}} u_b/H$
0.5	1900	124	$256 \times 97 \times 97$	12.2	2.0	31083
0.5	2000	142	$256 \times 97 \times 97$	13.9	2.3	21518
0.5	2205	154	$256 \times 97 \times 97$	15.2	2.5	6456
0.5	3000	203	$384 \times 97 \times 97$	13.3	3.3	8301
0.5	5000	325	$512 \times 129 \times 129$	16	4.0	12021
1.5	2205	147	$256 \times 97 \times 289$	14.4	2.4	18166
2	1080	74	$256 \times 97 \times 385$	7.3	1.2	8587
2	1100	75	$256 \times 97 \times 385$	7.4	1.2	11449
2	1150	77	$256 \times 97 \times 385$	7.6	1.3	11449
2	1200	81	$256 \times 97 \times 385$	7.9	1.3	12403
2	1250	84	$256 \times 97 \times 385$	8.3	1.4	12403
2	1300	89	$256 \times 97 \times 385$	8.7	1.4	12403
2	1350	92	$256 \times 97 \times 385$	9.0	1.5	8587
2	1400	95	$256 \times 97 \times 385$	9.4	1.6	10494
2	1450	99	$256 \times 97 \times 385$	9.7	1.6	8587
2	1500	102	$256 \times 97 \times 385$	10.0	1.7	13357
2	1600	108	$256 \times 97 \times 385$	10.6	1.8	8587
2	1800	121	$256 \times 97 \times 385$	11.9	2.0	8587
2	2205	145	$256 \times 97 \times 385$	14.2	2.4	23778
2	3000	191	$384 \times 97 \times 385$	12.5	3.1	9817
2	5000	297	$512 \times 129 \times 513$	14.6	3.7	6456
4	2205	143	$256 \times 97 \times 769$	14.0	2.3	11270
8	2205	143	$256 \times 97 \times 1153$	11.9	3.1	8140

Table 4.1: Parameters in the open duct simulations performed for this study.

A	Re_b	Re_τ	$M_x \times M_y \times M_z$	Δx^+	$\max\{\Delta y^+, \Delta z^+\}$	$t_{\text{stat}} u_b/h$
2	2205	143	$192 \times 129 \times 257$	9.4	3.5	12165
4	2205	142	$192 \times 129 \times 513$	9.3	3.5	17961
8	2205	142	$192 \times 129 \times 1025$	9.3	3.5	12060

Table 4.2: Parameters in the closed duct simulations performed for this study.

4.2 Simulation set-up

In total 23 open duct simulations with aspect ratio ranging from 0.5 to 8, and Reynolds number from 1080 to 5000 were performed until their statistical convergence for this study. Additionally, 3 closed duct simulations with aspect ratio $A = [2,4,8]$ were also performed for comparison purpose. The same fixed streamwise period of $L_x/H = 8\pi$ for the open duct simulations, and $L_x/h = 4\pi$ for the closed duct simulations were chosen as in the previous chapter.

All simulations presented here satisfy the following resolution requirements: CFL number is below 0.3; Δx^+ is below 16; the number of Chebyshev mode ensures that the maximum cross-sectional grid spacings $\max\{\Delta y^+\}$ and $\max\{\Delta z^+\}$ are below 4. Their statistics were accumulated over at least $6456H/u_b$ in order to achieve satisfactory convergence. Numerical parameters of those simulations are summarised in Table 4.1 and 4.2 for the open and closed duct simulations respectively.

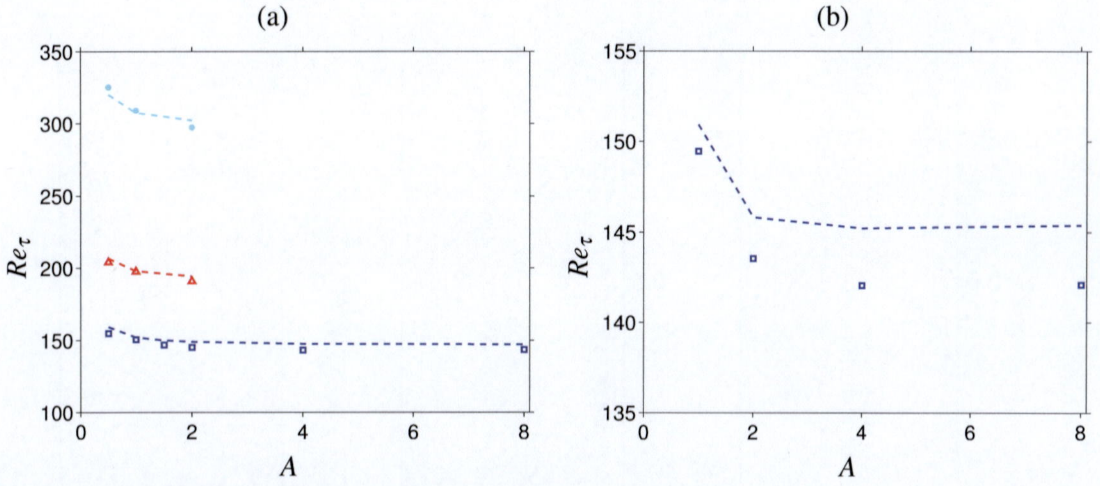

Figure 4.1: Friction-based Reynolds number Re_τ as function of aspect ratio A, of (a) open and (b) closed duct respectively. Coloured symbols indicate the value of the bulk Reynolds numbers: □, $Re_b = 2205$; △, $Re_b = 3000$; $*$, $Re_b = 5000$. Dashed lines indicates predictions by the empirical formula of Jones (1976).

Finally, it is worth noting that the empirical formula by Jones (1976) was confirmed to be valid to estimate the friction factor in both open and closed duct flow with the aspect ratio larger than unity (cf. Figure 4.1). The general form of the empirical formula is listed here for completeness:

$$f^{-1/2} = 2\log_{10}\left(\phi^* \frac{4A}{1+A} Re_b\, f^{1/2}\right) - 0.8 \tag{4.1}$$

where friction factor f is given by $f = 8u_\tau^2/u_b^2$, and shape factor ϕ^* that is a function of aspect ratio and can be approximated as:

$$\phi^* \approx \frac{2}{3} + \frac{11}{24}(AH)^{-1}(2-(AH)^{-1})\,. \tag{4.2}$$

Note that the duct full-height H needs to be replaced with the duct half-height h in the case of closed duct flow.

4.3 Aspect ratio dependence in marginally turbulent flow

We shall start this result section by investigating the influence of the aspect ratio on the marginal open duct turbulent flow. Two different aspect ratios of $A = [0.5, 2]$ were chosen for this purpose and comparisons with the reference marginal open duct turbulent flow with $A = 1$ and $Re_b = 1450$ ($Re_\tau \approx 102$) will be made from the viewpoint of the minimal set of coherent structures confined inside the duct cross-section. The procedure and criteria to define the minimum sustaining Reynolds numbers are same as the one explained in Chapter 3, which was originally employed in Uhlmann et al. (2007). The corresponding Reynolds number step sizes are again between $20-50$ bulk units.

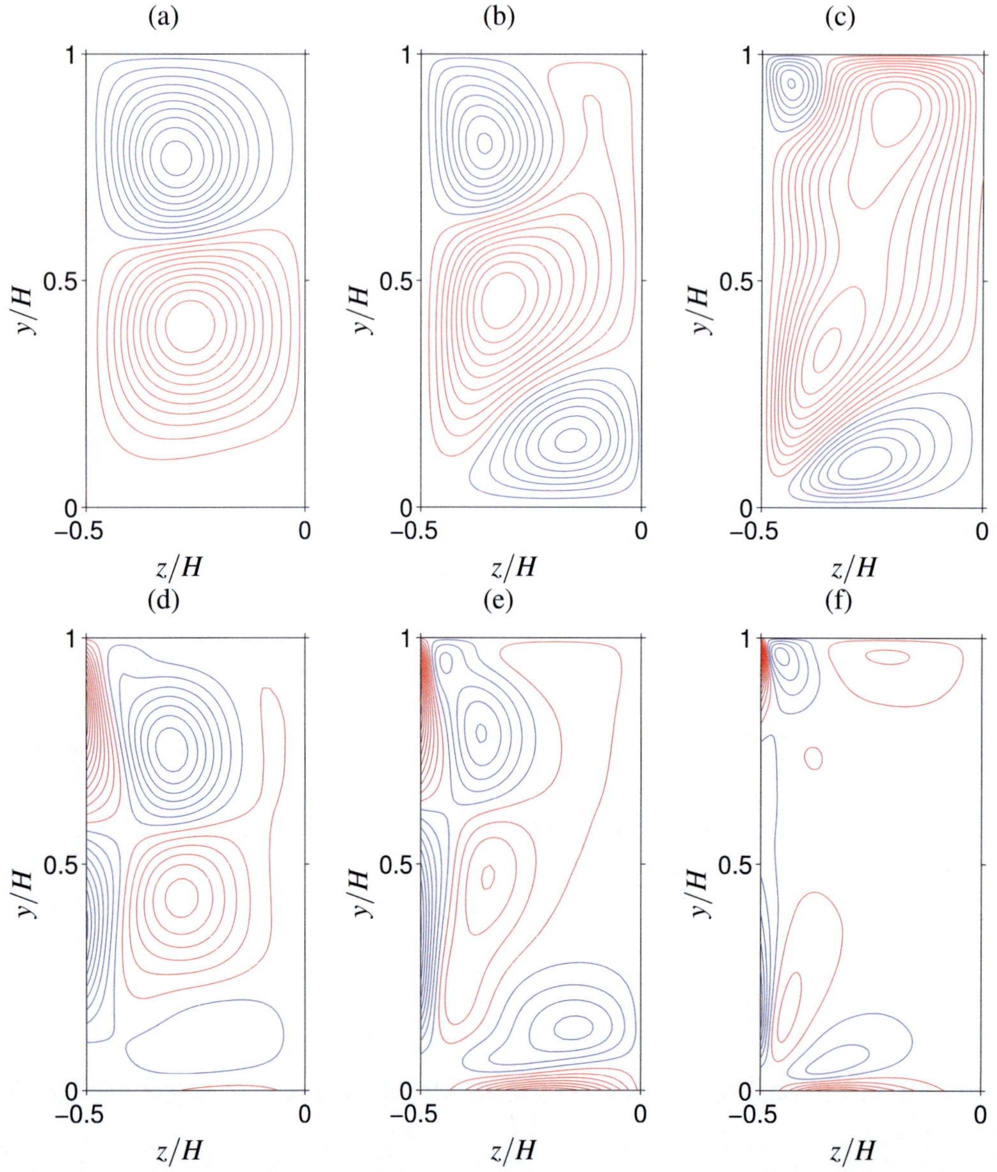

Figure 4.2: (a-c) Mean secondary flow streamfunction contour $\langle\psi\rangle$; (d-f) mean streamwise vorticity contour $\langle\omega_x\rangle$ with $[\max\langle\cdot\rangle - \min\langle\cdot\rangle]/20$. (a,d) $Re_b = 1900$; (b,e) $Re_b = 3000$; (c,f) $Re_b = 5000$ with $A = 0.5$. For both quantities, red and blue lines represent *positive* (clockwise) and *negative* (counter-clockwise) rotational motions respectively. Reflective symmetry about duct mid-span is applied and only left-half of the duct cross-section ($[0,H] \times [-AH,0]$) is shown.

4.3.1 Open duct flow with $A = 0.5$

On a course of successive Reynolds number reduction, the mean secondary flow pattern exhibits three-step qualitative transformation:

(i) eight-vortex mode exists with sufficiently high Reynolds number including the inner and outer secondary flow cells, the bottom-wall mean vortex and the side-wall mean vortex that was not observed in the flow with $A = 1$ (cf. Figure 4.2(c), and also Figure 3.7(b)); (ii) the inner secondary flow cell increases its diameter as Re_b decreases, and the streamfunction peak of the outer secondary flow cell disappears from the free-slip plane region consequently (cf. Figure 4.2(b));

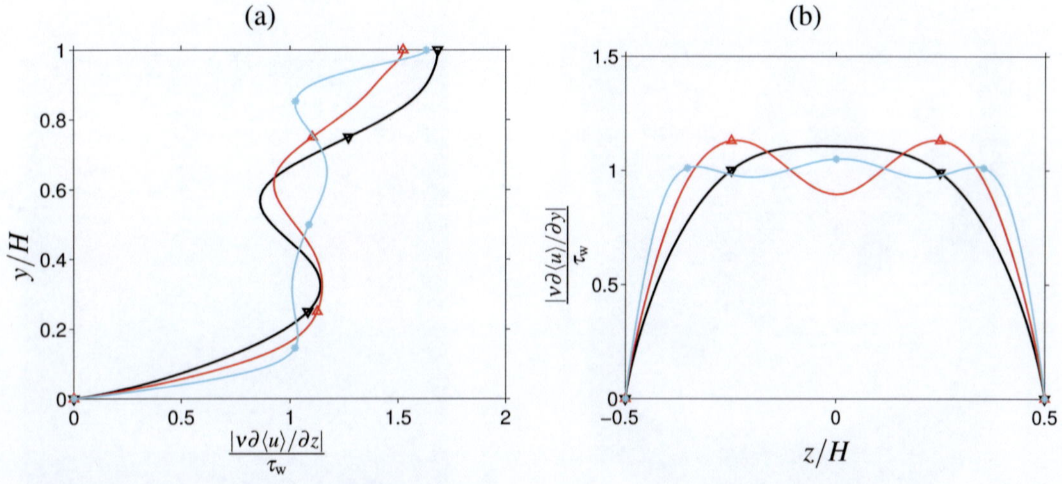

Figure 4.3: Mean wall stress of open duct flow with $A = 0.5$, distributions along: (a) side-wall and (b) bottom-wall, normalised by the whole no-slip wall average value τ_w. Coloured symbols indicate the value of the bulk Reynolds numbers: $\bigtriangledown$, $Re_b = 1900$; $\triangle$, $Re_b = 3000$; $*$, $Re_b = 5000$.

(iii) the bottom-wall streamfunction peak disappears from the duct cross-section, resulting a four-vortex state with a signature of one low-velocity streak on each duct side-wall, whilst the local wall stress minima disappears from the bottom-wall (cf. Figure 4.2(a) and Figure 4.3(b)).

The disappearance of the outer secondary flow in the state (ii) is a consequence of the inner secondary vortex, whose diameter scales in wall units, occupying the near free-slip plane region and any other types of mean vortices cannot be existed there. If the flow Reynolds number is sufficiently high (i.e. in the state (i)), however, the dimension of the inner secondary flow becomes small enough to allow the outer secondary flow to form in the region.

The existence of the very weak streamwise vorticity peak in the state (iii) above the bottom-wall boundary (cf. Figure 4.2(b)) should be also noted, which is absent from the secondary velocity profile. Moreover, the velocity dip-phenomenon on the duct mid-span occurs only in the state (i) (cf. Section 4.4.3 for more detailed discussions on this topic).

Now let us concentrate on the transition process to, and the four-vortex state itself. The local wall-stress minima (i.e. low-velocity streak signatures) on the bottom-wall disappear somewhere between $2000 \leq Re_b \leq 2205$ ($142 \leq Re_\tau \leq 155$), where the bottom-/side-wall edge length in wall units coincides well with the equivalent wall length in the minimal Reynolds number limit of the square closed duct flow ($2h^+ \approx 154$). Eventually the flow Reynolds number reaches to its lowest sustainable limit, which was found to be at $Re_b = 1900$ ($Re_\tau \approx 135$). Geometrically speaking, the minimal Reynolds number duct full-height in wall units ($H^+ \approx 135$) is approximately 20^+ shorter than the corresponding length in the minimal square closed duct flow.

As a result, the side-walls become predominantly "*active*" in terms of turbulent activities in comparison with the bottom-wall where it remains relatively quiescent. Instantaneous snapshot of turbulent coherent structures as well as the mean turbulent kinetic energy distribution shown in Figure 4.4 highlight the two-sided (i.e. left and right side-walls) existence of turbulence in the open duct four-vortex state well.

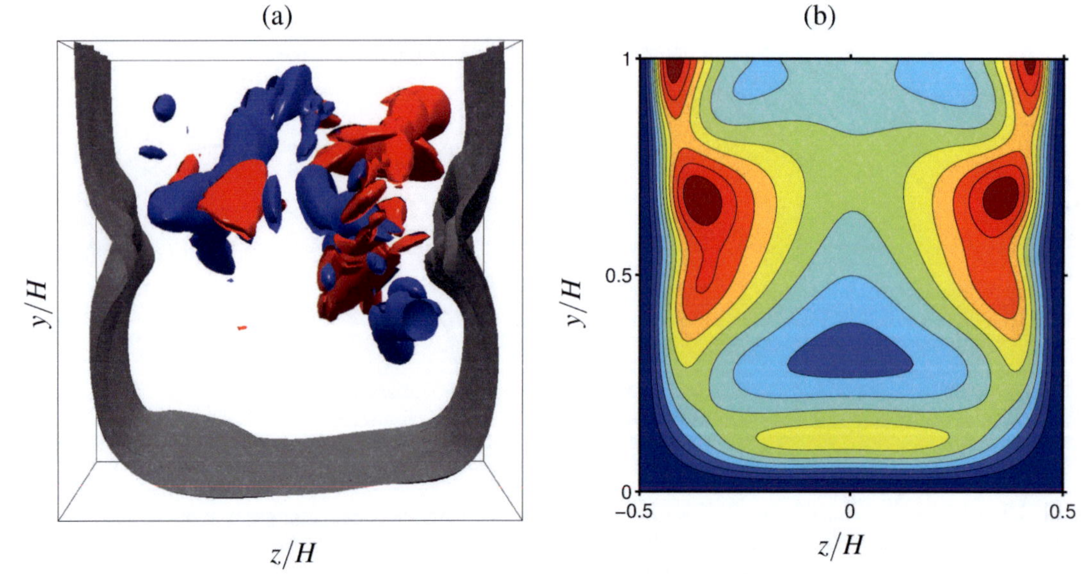

Figure 4.4: (a) Snapshot of instantaneous flow structure at $Re_b = 1900$ with $A = 0.5$. The streamwise velocity is shown by the gray iso-surface at the level of $0.6u_b$, and the streamwise coherent vortices are shown by the iso-surfaces of the second invariant of velocity gradient tensor (Q-criterion) at the level of $0.02u_\tau^4/\nu^2$. Colours of the structures indicate sign of their streamwise vorticity: red, *positive*; blue, *negative*. (b) Mean turbulent kinetic energy contour $0.5(u'^2 + v'^2 + w'^2)/u_\tau^2$, given by 0.3797(0.3797)3.797.

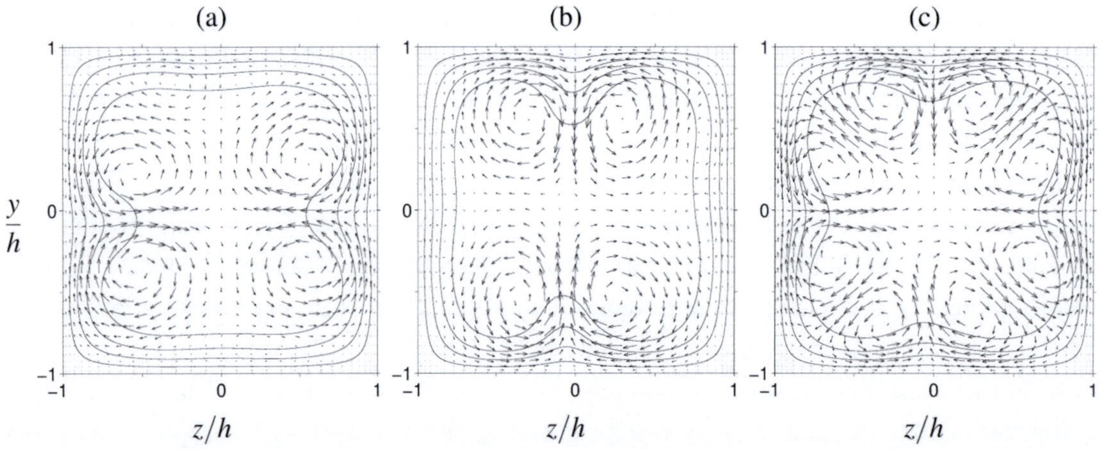

Figure 4.5: Contour lines of the primary mean flow $\langle u \rangle$ (with increment $\max\langle u \rangle/5$) and vectors of the secondary mean flow $\langle v \rangle$, $\langle w \rangle$ for the square closed duct with $Re_b = 1205$ and $L_x/h = 2\pi$: (a) averaging interval $771h/u_b$ in the side-wall dominant state; (b) a different interval with length $482h/u_b$ in the bottom-wall dominant state; (c) long-time integration including both former intervals ($1639h/u_b$). Vectors are shown for every third grid point. These plots are from Uhlmann et al. (2007).

The resulting mean flow pattern shows a great qualitative agreement with the alternating four-vortex pattern found in the square closed duct flow by Uhlmann et al. (2007) (note: their secondary flow profiles in a marginal square duct turbulence are reproduced in Figure 4.5 for comparison), though with one major difference that the open-duct four-vortex state does not exhibit any temporal-alternation of orientation such as the one observed in the closed duct counterpart.
In order to quantify the absence of such alternation, examining a temporal evolution of the indicator function introduced in the previous chapter (cf. Equation 3.3) is again useful. Figure 4.6 shows

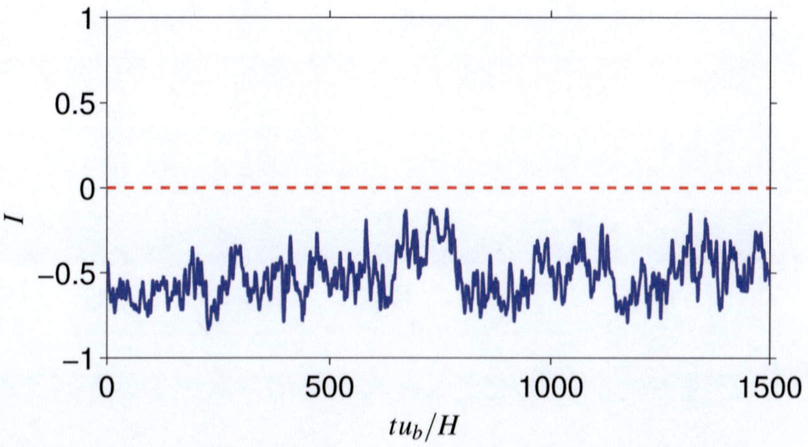

Figure 4.6: Temporal evolution of the indicator function I of the open duct simulation with $Re_b = 1900$ and $A = 0.5$. Red dashed line indicates $I = 0$.

an example of such temporal evolution of the indicator function in the open duct simulation with $Re_b = 1900$ and $A = 0.5$.

It is readily visible that the indicator value remains consistently around -0.5 without any signatures of clear step-like transition. Note that the average negative value confirms the side-wall dominance of the turbulent activities, since the definition of the indicator function is designed to yield zero, positive and negative average values when the near-wall turbulent activities are balanced, bottom-wall dominant and side-wall dominant respectively. Note also the consistency between the level of negative mean value (≈ -0.5) and the corresponding level in the minimal square closed duct flow during the side-wall active periods shown in Uhlmann et al. (2007) (cf. their Figure 4). Moreover, the absence of the alternation can be also observed directly from the temporal evolutions of the wall-stress distribution and the locations of the stress minima, as shown in Figure 4.7.

The clear contrast between the consistent existence of the side-wall low velocity streak around the side-wall bisector (detected as the local wall-stress minima), and the sparse appearance of the stress minima along the bottom-wall edge should also be noted.

A cause of such discrepancy, with respect to the temporal alternation observed in the closed duct minimal flow, is considered to be the broken reflective symmetry about the duct bisector in y-direction due to the presence of the free-slip plane. In other words, since turbulent streamwise vortices can approach closer to the free-slip boundary than to the no-slip walls (cf. Chapter 3), those side-wall structures can be sustained on a *shorter* edge in wall units than the one for the bottom-wall structures without being relaminarised, in spite of the fact that the extensions of side-wall and bottom-wall edges are identical in this particular aspect ratio (i.e. $L_y = L_z = H$). Consequently, the predominance of the side-wall turbulent activities is a stable state in the open duct marginal turbulence, therefore no alternations of the orientation are observed.

This open duct four-vortex state is a remarkable phenomenon in a sense that, in this narrow aspect ratio configuration, the marginal open duct turbulent flow behaves similar to the full square duct marginal turbulence, in contrast to the marginal turbulence in the open duct with $A = 1$, where the mean secondary flow pattern shows a strong similarity to a semi-cross-sectional mean profile of the

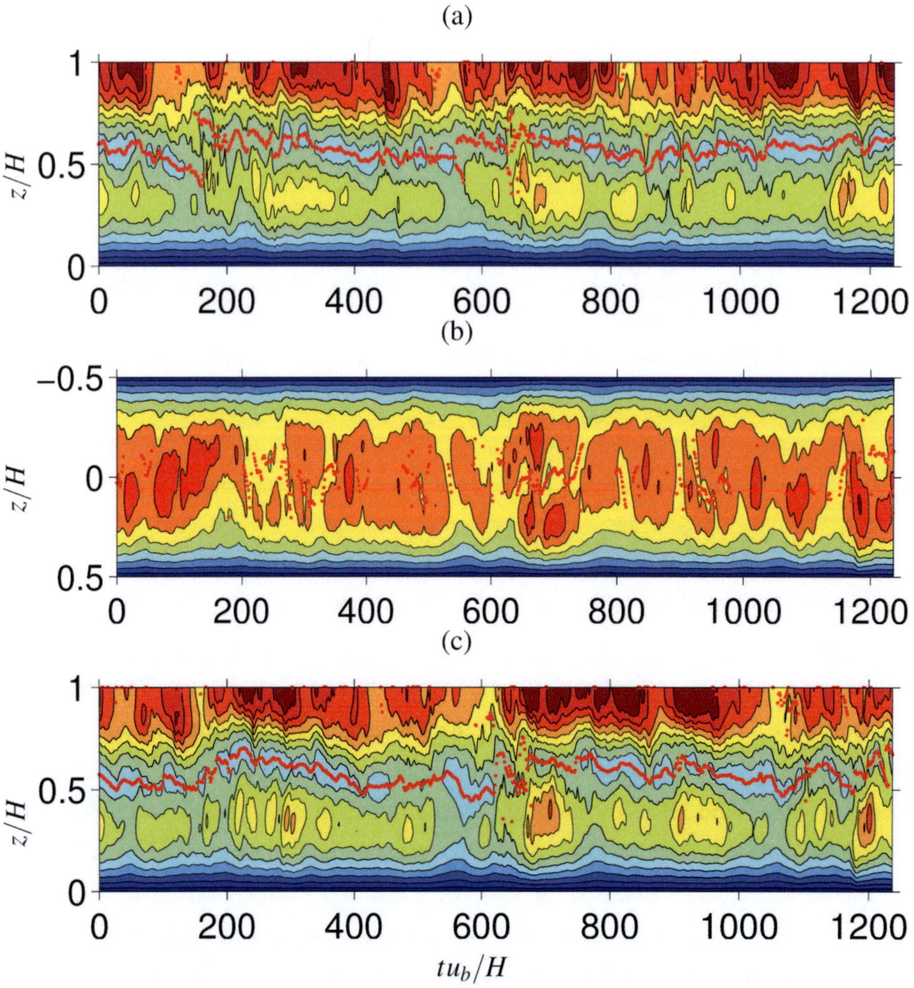

Figure 4.7: Temporal evolution of: (a) left and (c) right side-wall stress distributions respectively; (b) distribution of bottom-wall stress, both averaged instantaneously over the full streamwise domain length, and plotted as functions of bulk time tu_b/H. In all cases, the red dots indicate the locations of wall-stress minima of each averaged snapshot. The contour lines indicate $0(0.2)2.2$ times the no-slip boundary averaged wall stress τ_w. Statistics are from 1000 instantaneous fields separated by $10\nu/u_\tau^2$ of open duct flow with $A = 0.5$ and $Re_b = 1900$.

fully-developed square closed duct turbulence (except in the small region in the mixed-boundary corners) instead (cf. Section 3.3.1). This qualitative difference can be explained by the higher degree of geometrical and dynamical constraints imposed upon the turbulent coherent structures in the spanwise direction in the narrow open duct flow. In such spanwise-restricted environment, the coherent structures are not given enough space to exploit the spanwise transport enhancement from the free-slip boundary condition, in contrast to the free-slip-plane normal direction where the boundary condition effect is noticeable in the mean locations of the structures shifted towards the free-slip plane. As a result, the marginal turbulence in the narrow open duct shows its mean flow profile more similar to the closed duct counterpart than the marginal turbulence exists in the open duct flow with larger aspect ratios, due to the restricted influence of the spanwise transport mechanism, which was found to be a key mechanism to create the unique mean flow patterns observed in the open duct turbulence, as discussed in the previous chapter.

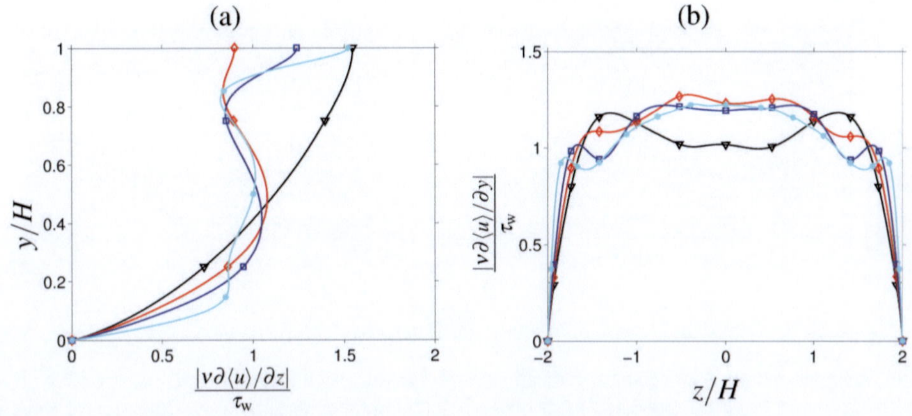

Figure 4.8: Mean wall stress of open duct flow with $A = 2$, distributions along: (a) side-wall and (b) bottom-wall, normalised by the whole no-slip wall average value τ_w. Coloured symbols indicate the value of the bulk Reynolds numbers: ∇, $Re_b = 1080$; $\triangle$, $Re_b = 1900$; $\square$, $Re_b = 2205$; $*$, $Re_b = 5000$.

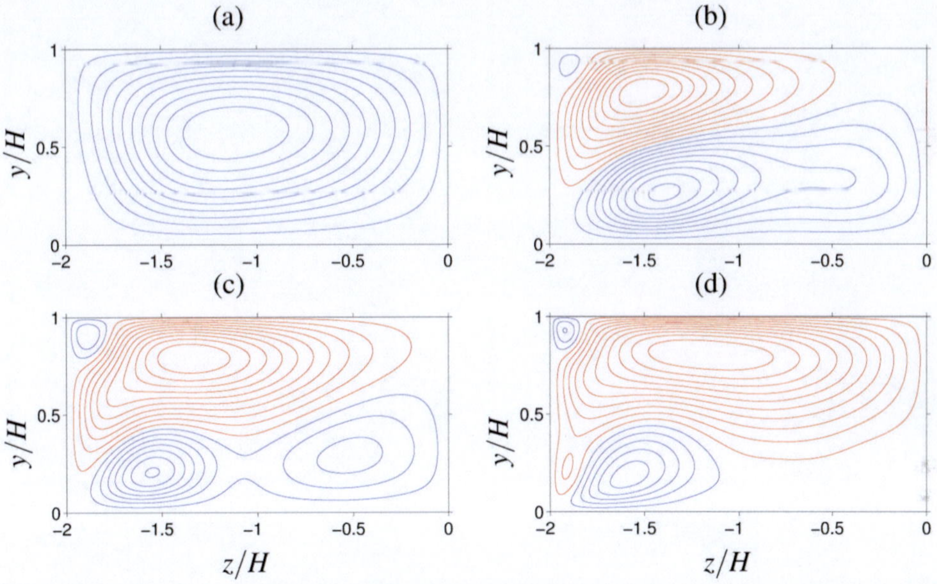

Figure 4.9: Streamlines $\langle\psi\rangle(y,z)$ of mean secondary flow computed using $\langle v\rangle$ and $\langle w\rangle$ with reflective symmetry about duct mid-span (with increment $[\max\langle\psi\rangle - \min\langle\psi\rangle]/20$). Red lines correspond to clockwise rotation and blue lines to counter-clockwise motion. (a) $Re_b = 1080$; (b) $Re_b = 1500$; (c) $Re_b = 2205$ (d) $Re_b = 5000$ with $A = 2$. Only left-half of the duct cross-section ($[0,H] \times [-AH,0]$) is shown.

4.3.2 Open duct flow with $A = 2$

Similar to the previous observation for $A = 0.5$, open duct flow with $A = 2$ also exhibit a drastic change of their secondary flow pattern in the limit of the minimal Reynolds number, which was found at $Re_b \approx 1080$ ($Re_\tau \approx 74$). However, the feature of the resulting secondary flow pattern is remarkably different: in contrast to the four-vortex state for $A = 0.5$ in which the bottom-wall low-velocity streaks become absent, the signature of the side-wall low-velocity streaks disappear from the side-wall stress distributions (cf. 4.8(a)) somewhere around $1150 \leq Re_b \leq 1200$ ($77 \lesssim Re_\tau \lesssim 81$) instead, and a two-vortex state appears in the secondary flow profile (cf. Figure 4.9(a) and 4.10).

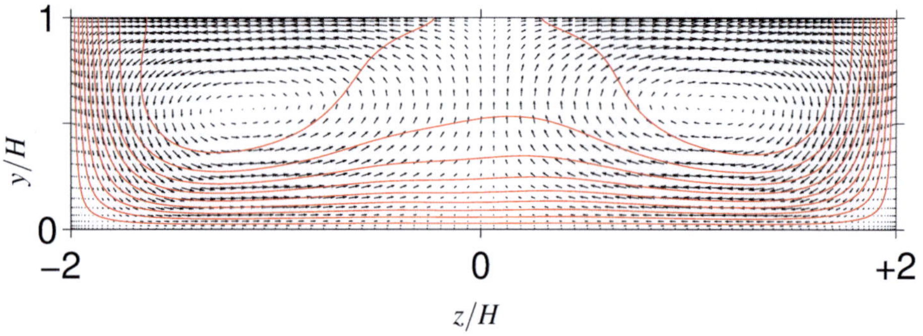

Figure 4.10: Mean secondary flow vector and mean streamwise velocity contour of open duct flow with $A = 2$ and $Re_b = 1080$. Red contour lines indicate 0.1(0.1)0.9 times the maximum value of the mean streamwise velocity $\langle u \rangle$.

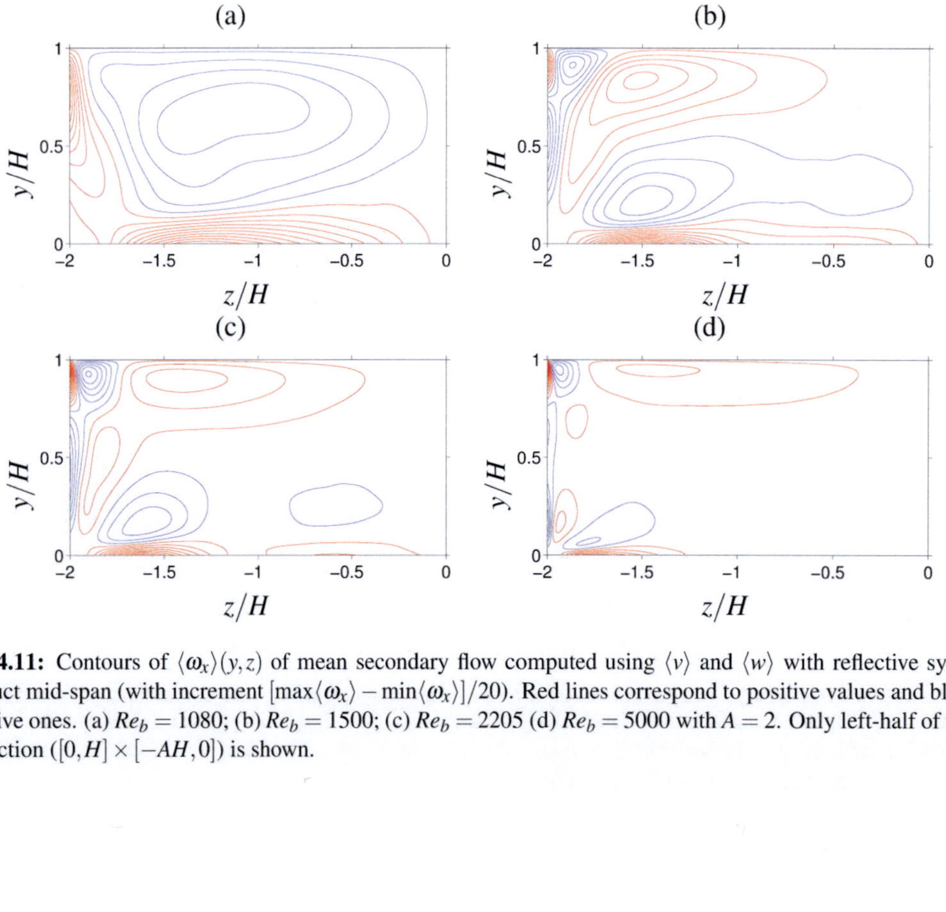

Figure 4.11: Contours of $\langle \omega_x \rangle (y,z)$ of mean secondary flow computed using $\langle v \rangle$ and $\langle w \rangle$ with reflective symmetry about duct mid-span (with increment $[\max\langle \omega_x \rangle - \min\langle \omega_x \rangle]/20$). Red lines correspond to positive values and blue lines to negative ones. (a) $Re_b = 1080$; (b) $Re_b = 1500$; (c) $Re_b = 2205$ (d) $Re_b = 5000$ with $A = 2$. Only left-half of the duct cross-section ($[0, H] \times [-AH, 0]$) is shown.

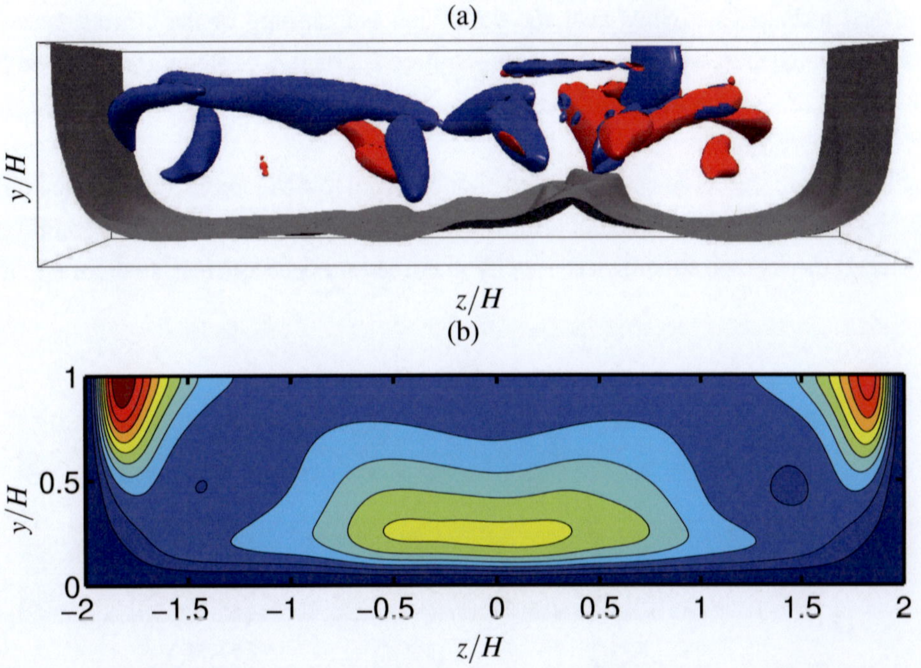

Figure 4.12: (a) Snapshot of instantaneous flow structure at $Re_b = 1080$ with $A = 2$. The streamwise velocity is shown by the gray iso-surface at the level of $0.6u_b$, and the streamwise coherent vortices are shown by the iso-surfaces of the second invariant of velocity gradient tensor (Q-criterion) at the level of $0.01u_\tau^4/\nu^2$. Colours of the structures indicate sign of their streamwise vorticity: red, *positive*; blue, *negative*. (b) Mean turbulent kinetic energy contour $0.5(u'^2 + v'^2 + w'^2)/u_\tau^2$, given by 0.8762(0.8762)8.762.

The remaining vortex pair is initially the one on the bottom-wall, grows its size as Reynolds number is lowered, and eventually occupies the entire duct cross section. Moreover, it is interesting to notice that the resulting mean two-vortex picture matches qualitatively to the averaged secondary flow pattern of the leading part of the open duct puff, reported by Nowakowski (2007) for $A = 1$. As far as the instantaneous coherent structures are concerned, the enlarged mean secondary flow vortex cells are the likely-footprints of a pair of near-wall vortices with opposite signs of rotation, flanking a low-velocity streak and drifting in spanwise direction as seen in Figure 4.12(a).

Such spanwise meandering motion of the structures is evident from the mean bottom-wall stress distribution shown in Figure 4.8(b), where a distinguishable plateau forms around the bottom-wall mid-span analogous to the wall stress profile for the higher Reynolds number flow, witnessing the existence of a translational symmetry along the boundary except in the vicinity of the solid-solid corners. Moreover, the spanwise meandering was observed to have a long period of cycle (cf. Figure 4.13 for a spatio-temporal evolution of the low-speed streak location represented as the wall stress distribution), and lead to a very long averaging period to achieve a satisfactory statistical convergence with $t_{\mathrm{stat}}u_b/H \gtrsim 9000$. Note that the absence of the side-wall low-velocity streaks can be confirmed in Figure 4.13(a,c).

At first glance, the resulting two-vortex state can be seen as a consequence of the bottom-wall being active whilst the side-wall turbulent activities become quiescent due to the local relaminalisation as before. However, the increased amplitude of the side-wall stress distributions in the two-vortex state somehow conflicts with our expectation, since such increase is usually associated with increased turbulent activities near the walls that are missing in the current case. Consequently, the observed side-wall stress increase requires alternative explanations, and we postulate that the locations of streamwise velocity maxima, shifted towards the side-wall boundaries shown in Figure 4.10, are relevant.

The velocity maxima shift is an outcome of the bottom-wall low-velocity streak located on the duct mid-span being lifted as high as the duct full-height, dividing the duct cross-section into two cells where each cell has its own streamwise velocity maxima as a consequence. In contrast, other flow

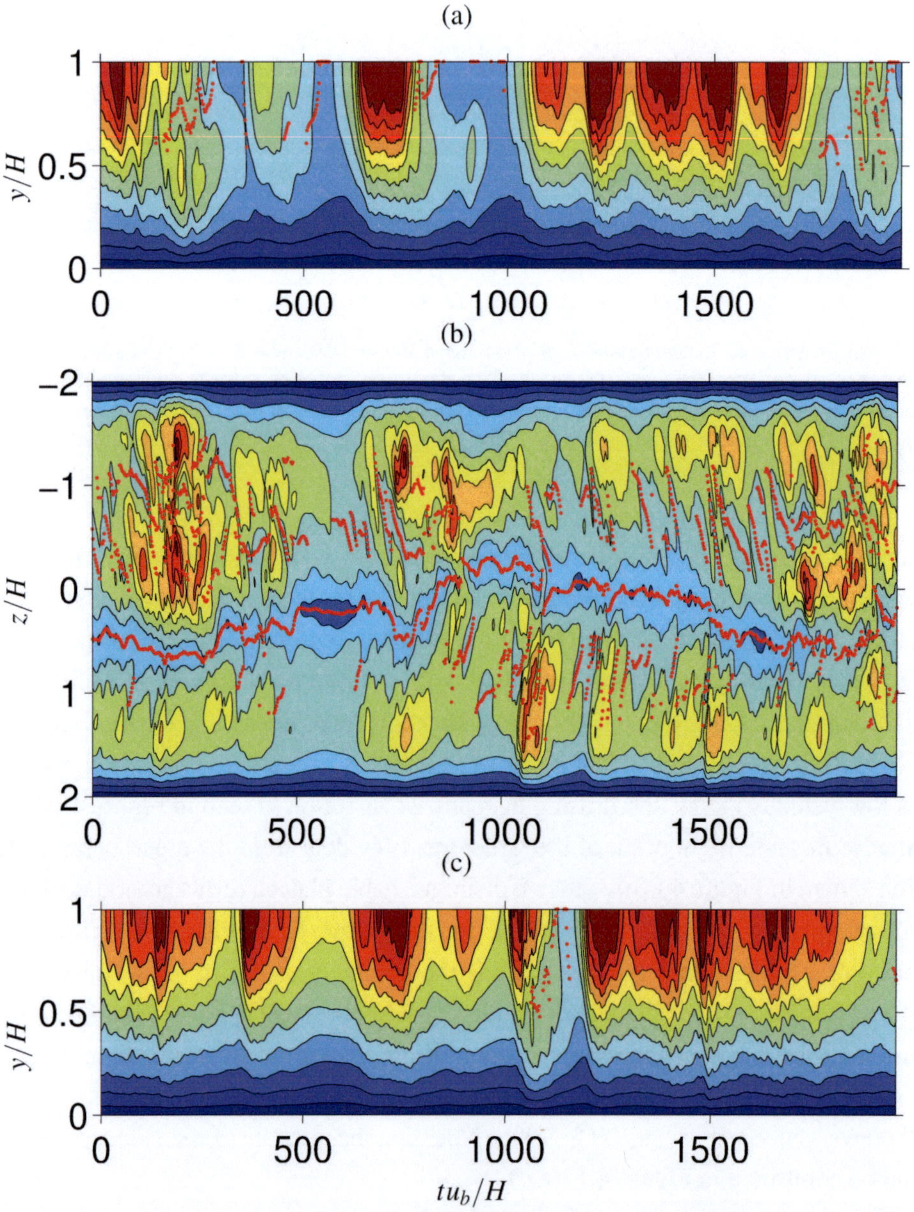

Figure 4.13: Temporal evolution of: (a) left and (c) right side-wall stress distributions respectively; (b) distribution of bottom-wall stress, both averaged instantaneously over the full streamwise domain length, and plotted as functions of bulk time tu_b/H. In all cases, the red dots indicate the locations of wall-stress minima of each averaged snapshot. The contour lines indicate 0(0.2)2.2 times the no-slip boundary averaged wall stress τ_w. Statistics are from 1000 instantaneous fields separated by $10\nu/u_\tau^2$ of open duct flow with $A = 2$ and $Re_b = 1080$.

configurations discussed so far only have one global velocity maximum, observed consistently on the duct mid-span. Through this mechanism based on the shifted streamwise velocity maxima, the duct side-wall regions are not necessarily required to be turbulent to exhibit the high amplitude of the side-wall stress.

An alternative explanation can also be postulated from the observation that the TKE profile, shown in Figure 4.12(b), accommodates two strong peaks in the mixed-boundary corners, indicating the existence of quasi-streamwise vortices in the corner regions. Based on the vortex dipole hypothesis discussed in the previous chapter, it is possible to conjecture that the vortices with certain sense of rotation generated around the duct mid-span are transported to the mixed-boundary corners. Note that those vortices cannot be autonomously regenerated near the mixed-boundary corners, since the low-velocity streaks necessary for the autogeneration are absent, but it does not automatically imply the absence of vortices in combination with the aforementioned transport mechanism. Further structural and dynamical investigations are required to confirm the validity of above explanations.

Now, let us investigate the local relaminalisation and consider a consequence of the related solo-existence of the low-velocity streak on the bottom-wall edge, although the friction-based spanwise extension (i.e. $L_z^+ = (2AH)^+ \approx 300$, in contrast to $A = 1$ case where the five-streak state first appears at $L_z^+ \approx 250$) should be sufficiently wide to accommodate two low-velocity streaks instead. A key ingredient here is the transitional closed duct turbulence study performed by Takeishi et al. (2015) for a range of aspect ratios $A = [1 \dots 9]$. They observed that the streamwise localised turbulent puff detached from the side-walls somewhere around $4 \leq A \leq 5$, where the *local* friction-based side-wall edge length (i.e. $Hu_{\tau s}/\nu$, where $u_{\tau s}$ is mean frictional velocity based on the side-wall stress averaged only over the side-wall boundaries) drops to around 75 when aspect ratio is increased, caused by a reduction of local streamwise velocity gradient in the wall-normal direction. As a result, the side-wall boundary in the local wall units becomes too short to accommodate the minimum set of the near-wall turbulent structures, therefore the local relaminalisation of the side-wall turbulence occurs consequently (cf. their Figure 5). Figure 4.14(a) shows the corresponding local side-wall length scale $Hu_{\tau s}\nu$ of our open duct simulations, and indicates that indeed the measure drops to around 75 when the two-vortex state appears. From such observation as well as the quiescent turbulent activities illustrated by a low value of turbulent kinetic energy in the bulk regions near the solid-solid corners (cf. Figure 4.12(b)), we conclude that the local relaminalisation does occur in the marginal open duct turbulence as well.

Consequently, the effective turbulent spanwise extension, which is the duct full-span *excluding* the side-wall quiescent regions, becomes narrower and the single low-velocity streak state (or three-streak state by taking into account the accompanying high-velocity streak pair) is observed for slightly higher Reynolds number limit than predicted.

To this end, it is important to investigate the existence of the temporal-alternation of orientation as before, by means of an indication function that is based on the integrated turbulent quantities on cross-sectional sub-domains. Related to the unique vortex arrangement observed in the current flow configuration, we consider the following three equiareal triangular sub-domains:

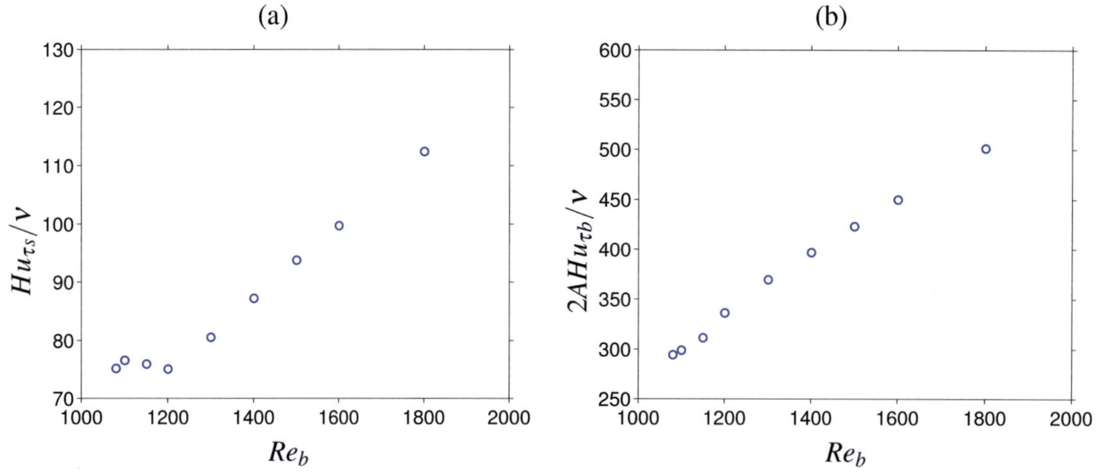

Figure 4.14: (a) Duct full-height H and (b) full-span as a function of bulk Reynolds number Re_b, normalised by the side-wall and bottom-wall friction length respectively. Open duct flow with $A = 2$.

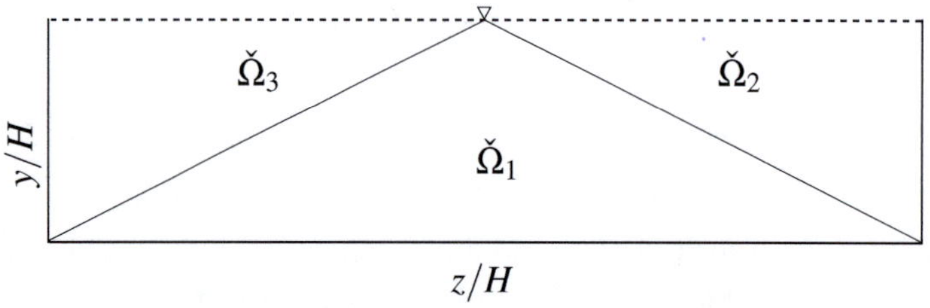

Figure 4.15: Schematic of alternative triangular sub-domain $\check{\Omega}_i$

$$\begin{cases} \check{\Omega}_1 \;:\; \{(y,z)|(y-H)/H < z/(AH) \cap (y-H)/H < -z/(AH)\}, \\ \check{\Omega}_2 \;:\; \{(y,z)|(y-H)/H < z/(AH) \cap (y-H)/H > -z/(AH)\}, \\ \check{\Omega}_3 \;:\; \{(y,z)|(y-H)/H > z/(AH) \cap (y-H)/H > -z/(AH)\}. \end{cases} \tag{4.3}$$

and the resulting arrangement is illustrated in Figure 4.15.

The corresponding indicator function is again based on the time-evolution of the streamwise-averaged squared streamwise vorticity integrated over one of the above three triangular sub-domains, defined as:

$$\check{I}(t) = \frac{\check{S}_1 - \check{S}_2 - \check{S}_3}{\check{S}_1 + \check{S}_2 + \check{S}_3}, \tag{4.4}$$

where

$$\check{S}_i = \int\!\!\int_{\check{\Omega}_i} \langle \omega_x \rangle_x^2 dydz\,. \tag{4.5}$$

Note that the new indication function is again bounded by $-1 \leq \check{I} \leq 1$, and designed to yield *positive* and *negative* values in case of bottom-wall and side-wall dominance of turbulent activities respectively.

An example of such temporal evolution of the new indication function for the current marginal flow is shown in Figure 4.16, together with the equivalent evolution of our standard indication function I. It is indeed visible from the evolution of the new indication function that there exists some

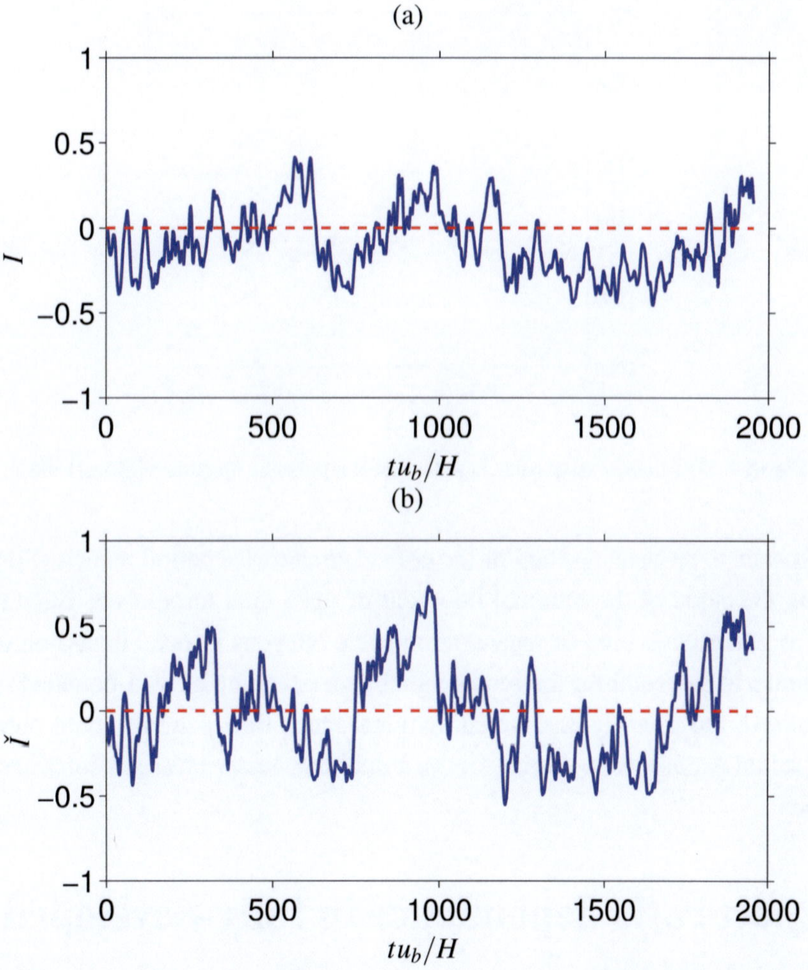

Figure 4.16: (a) Temporal evolution of the indicator function I of the open duct simulation with $Re_b = 1080$ and $A = 2$. (b) same plot with the alternative indicator function $\check{I}$. Red dashed lines indicate $I = \check{I} = 0$.

sort of alternation between the bottom-wall and side-wall dominance of the turbulent activities captured by the integrated squared streamwise vorticity. Further investigations are required for a structural explanation of such alternation, and for the discrepancy with respect to the low-velocity streak absence on the duct side-walls demonstrated earlier.

To conclude above discussions on the marginally turbulent open duct flow with different aspect ratio A, it is important to notice that the mean secondary flow vortices in the mixed-boundary corners are always rotating in the direction of the inner secondary flow, irrelevant to any particular vortex-mode (i.e. four-, six-, or two-vortex mode). We associate this phenomenological consistency with the sorting mechanism driven by the vortex-free-surface interactions discussed in the previous chapter, in a sense that the unfavoured vortices cannot exist stably in the mixed-boundary corners to appear in the mean flow statistics. Secondly, aspect ratio dependence in the minimal sustainable Reynolds number illustrated in Figure 4.17 needs to be noted. Such dependence coincides qualitatively with the observations of Takeishi et al. (2015) for the closed duct transitional turbulence (cf. their Figure 1), and can be consistently explained as a consequence of weakening constraint by the side-walls. Finally, it is highly desirable to determine the critical aspect ratio for

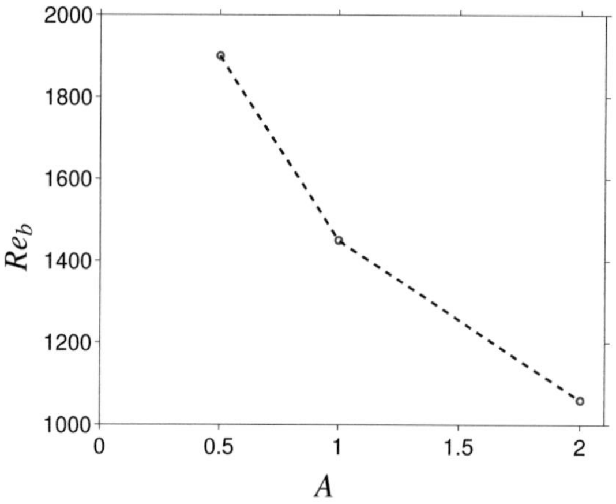

Figure 4.17: Minimal sustainable bulk Reynolds number as function of aspect ratio A.

the two-vortex state to be sustained and/or the critical streamwise period, which will pave the way for completing the study of the minimal flow unit of open duct turbulence. Such minimal flow are expected to underline a kind of regeneration cycle happens in open ducts: an essential kind of material known to be useful for the fundamental understanding of wall-bounded turbulence. In the similar context, those newly-discovered marginal mean flow patterns could potentially serve as promising initial conditions to search for open duct non-linear invariant solutions that are still to be discovered.

4.4 Aspect ratio dependence in fully-developed turbulent flow

Let us now consider the cases of fully-developed open duct turbulent flow in different aspect ratios. For this purpose, we simulated the open duct flow with the fixed Reynolds number at $Re_b = 2205$ and with six different aspect ratios ($A = [0.5, 1, 2, 3, 4, 8]$). Additionally, four accompanying closed duct simulations were performed with the same Reynolds number $Re_b = 2205$ and with $A = [1, 2, 4, 8]$.
We will investigate their mean flow statistics in the following manner. Firstly, a consequence of the near-wall velocity streaks will be deduced from the distribution of the wall stress, and discussed from the viewpoint of the geometrical constraint on the integer number of structures existing on the boundaries. Secondly, the mean secondary flow patterns will be discussed in terms of the cross-sectional streamfunction and the streamwise vorticity peak locations, which will be quantified as before. Such quantifications will help us to verify the early experimental observations of Tominaga et al. (1989) from the viewpoint of the mean flow geometries that the spanwise extension of the outer secondary flow reaches up to $2H$ in the case of high aspect ratio duct flow, whereas the mean bottom-wall vortices extend only up to H in spanwise direction. Thirdly, spanwise distribution of the secondary flow intensity integrated in the free-slip plane normal direction will be examined, with a special emphasis upon the decay properties of the quantity away from the side-walls. Such investigations of the integral quantities should provide us further supports on the confinement of

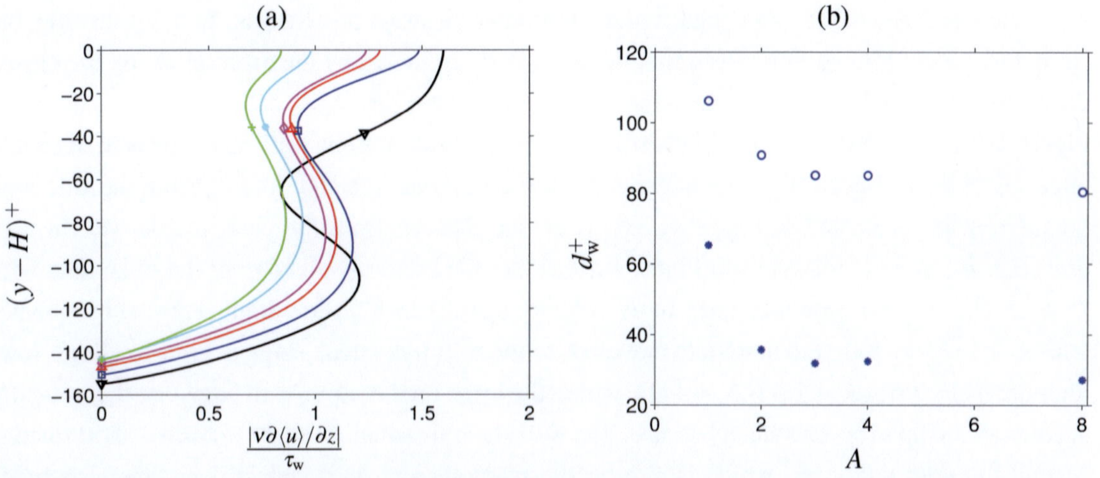

Figure 4.18: (a) Mean side-wall stress distribution normalised by the average over the whole no-slip wall τ_w. The origin is shifted to the mixed-boundary corner and the distance is presented in wall units. Reynolds number is fixed at $Re_b = 2205$ and coloured symbol code is as of Figure 4.19. (b) Distance to the prescribed corners d_w^+, in wall units, of the side-wall stress maxima and minima as a function of Re_τ. Here $*$ is the position of minima next to the free-slip plane, while $\circ$ represent the location of the maxima directly below.

the secondary flow within the side-wall regions. Finally, investigation on the aspect ratio and Reynolds number dependence of the streamwise velocity dip-phenomenon will conclude this part of the current study.

4.4.1 Mean streamwise structure

Let us start this discussion with a consequence of the side-wall stress distributions, which is unique to the open duct flow. Again, we associate the preferential locations of high- and low-velocity streaks with the local maxima and minima of the stress distribution, and their distance to the free-slip plane will be quantified by means of one-dimensional Newton's method.

Figure 4.18(a) shows the mean side-wall stress distributions in the open duct flow, whereas Figure 4.18(b) shows the corresponding location of the first local minimum of mean wall-shear and the following local maximum relative to the free-slip plane.

Firstly, it can be seen that a local maximum of wall-shear is always located directly on the free-slip plane, as it was shown for the open duct with $A = 1$. Secondly, the stress distribution in the open duct with $A = 0.5$ shows a profile similar to the one from the marginal open duct flow investigated in the previous chapter, accommodating a low-velocity streak slightly above the side-wall bisector. Subsequently, the velocity streak locations move towards the free-slip plane as aspect ratio increases, exhibiting a trend analogous to the open duct flow with $A = 1$ whose flow Reynolds number being reduced successively. As a particular example, the low- and high-velocity streak locations from the free-slip plane for the largest aspect ratio ($A = 8$) are approximately at $30\delta_v$ and $80\delta_v$ respectively. Such velocity streak locations are approximately equivalent to the open duct with $A = 1$ at $Re_b = 1800$, presented in Figure 3.5(b). The above observation is consistent with the earlier discussion on the reduced local friction-based Reynolds number in the case of increased aspect ratio, first observed by Takeishi et al. (2015).

Let us now proceed to the consequence of the bottom-wall stress distribution. In the following, the open duct results will be introduced first, which will be followed by the discussions on the closed duct flow, compared accordingly with the open duct counterpart.

Figure 4.19(a,b) show the spanwise distributions of the mean bottom-wall shear stress in open and closed duct flow respectively. The locations of those first maxima and minima from the side-wall boundaries are presented in Figure 4.19(c,d) as the distance from the nearest side-wall in wall units d_w^+. It can be observed from Figure 4.19(a) that the bottom-wall edge of the open duct with $A = 0.5$ ($L_z^+ \approx 154$) accommodates three velocity streaks in high-low-high order, subsequently with $A = 1$ ($L_z^+ \approx 300$), the flow state moves on to the next five-streak stage in high-low-high-low-high ordering. With $A = 1.5$ ($L_z^+ \approx 441$), where the open duct bottom-wall edge can theoretically accommodate up to seven velocity streaks, the wall shear distribution starts to form a short plateau around the duct mid-span, which is similar the phenomenon observed in the higher Reynolds number flow at $A = 1$. At this stage, the geometrical constraint from the side-walls starts to become less significant in the duct mid-span region, and the velocity streaks in the region starts to gain some freedom of movement. As a result, the aforementioned plateau appears in the mean stress profile, instead of seven distinguishable peaks.

Subsequently, the duct mid-span plateau continues to occupy the central region on the bottom-wall boundary as the duct aspect ratio increases to 2, 4 and 8, whereas the locations of those stress maxima and minima in the near side-wall regions remain unchanged relative to the side-walls. Such observation was quantified in Figure 4.19(c), where the side-wall distance of the high-velocity streak peaks is shown to stay consistently around $50\delta_v$ away from the side-wall boundaries, whereas the neighbouring low-velocity streak peaks reside around $90\delta_v$ away from the same boundaries, being independent of the aspect ratio for $A \geq 2$.

A qualitative difference in the bottom-wall low-velocity streak locations should be noted in comparison with the result of the similar analysis performed in the previous chapter, where the low-velocity streak distance from the nearest side-walls was shown to first increase then decrease as the flow Reynolds number increases in the open duct with fixed aspect ratio at unity (cf. Figure 3.4(c)). Such comparison is interesting from the point of view that increasing Reynolds number with a fixed aspect ratio and increasing aspect ratio with a fixed Reynolds number are equivalent in terms of increasing the bottom-wall edge length in wall units.

Some quantitative differences can be also observed, for instance, with the open duct flow with $A = 1.5$ ($L_z^+ \approx 441$) which is equivalent in terms of the bottom-wall edge length in wall units to the flow with $Re_b = 3500$ at $A = 1$, whose corresponding normalised edge length is 452. In spite of such similarity, their low-velocity streak distance from the side-walls differ significantly ($d_\mathrm{w}^+ \approx 88$ and ≈ 104 respectively). Similarly, the open duct flow with $A = 2$ ($L_z^+ \approx 580$) is equivalent to the flow with $Re_b = 4500$ at $A = 1$, which has the corresponding bottom-wall length in wall units of 564. Their low-velocity streak distance from the side-walls are $d_\mathrm{w}^+ \approx 89$ and ≈ 113 respectively. It is interesting to observe that in both comparisons, the low-velocity streaks in the open ducts with larger aspect ratios remain closer to the side-walls, if they are compared with the corresponding streaks in the open duct with $A = 1$.

Let us now turn our attention to the bottom-wall stress distribution of the accompanied closed duct flow with the corresponding aspect ratios shown in Figure 4.19(b).

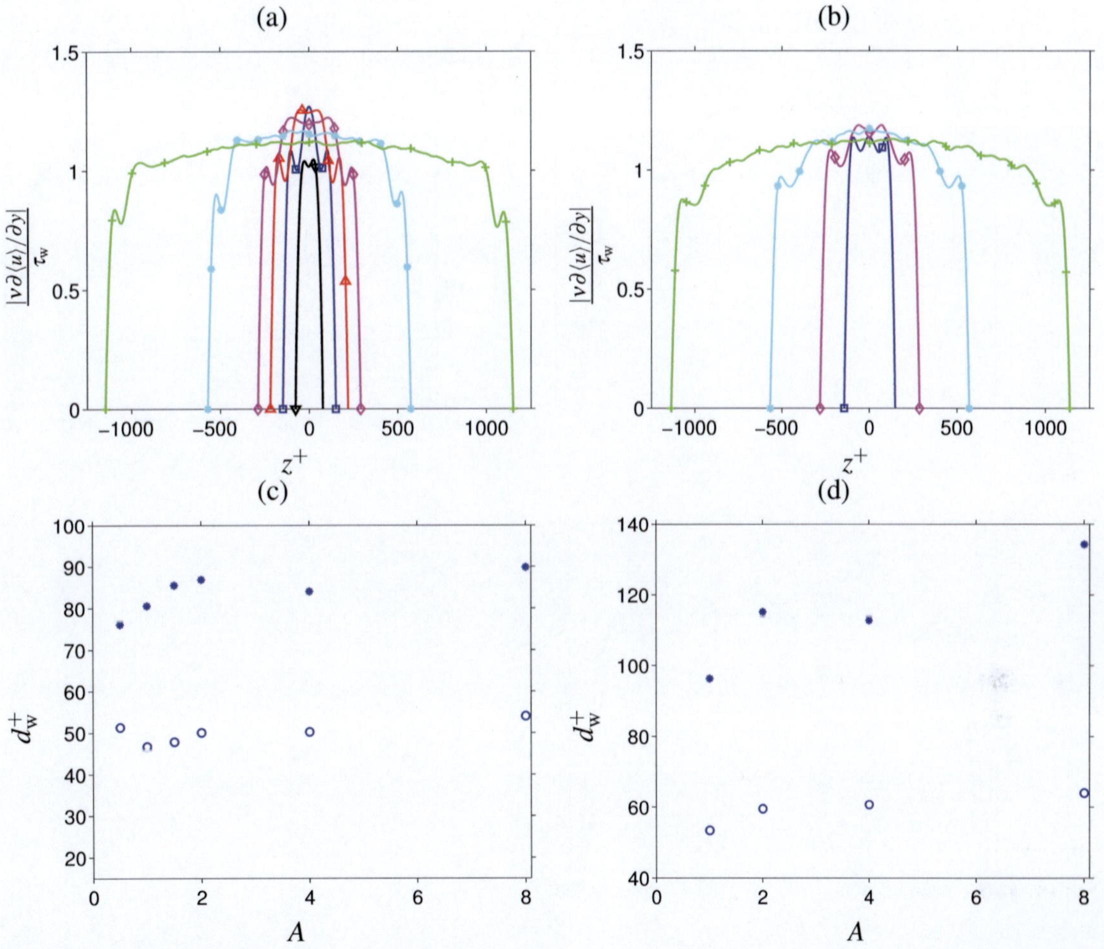

Figure 4.19: (a,b) Mean bottom-wall stress distributions of: (a) open and (b) closed ducts, normalised by the whole no-slip wall average value τ_w and presented as a function of z^+. Reynolds number is fixed at $Re_b = 2205$ and coloured symbols indicate different aspect ratio: ∇, $A = 0.5$; $\square$, $A = 1$; $\triangle$, $A = 1.5$; $\Diamond$, $A = 2$; $*$, $A = 4$; $+$, $A = 8$. (c,d) Distance to the closest corner d_w^+, in wall units, of the location of the near-corner maxima and minima of bottom-wall shear as a function of Re_τ. (c) open and (d) closed ducts. Here symbols $\circ$ and $*$ represent the position of the shear maxima and minima respectively.

As in the open duct flow, the closed duct flow with $A = 1$ exhibits a five-streak state, then the bottom-wall stress profile develops a plateau in the duct mid-span region with $A \geq 2$.
The aspect ratio dependence of the corresponding high- and low-velocity streak locations relative to the side-walls also show a trend similar to the open-duct counterpart: keeping approximately constant distances of $d_w^+ \approx 60$ and 110 respectively for $A \geq 2$. Note that the radical change in the relative location of the high-velocity streak in the closed duct flow with $A = 8$ should be a consequence of the the marginal statistical convergence evident from the less degree of symmetry about the duct mid-span.
The closed duct flow with $A = 2$ has the normalised bottom-wall edge length ($L_z^+ \approx 286$) equivalent to the one of the square closed duct flow with $Re_b = 4500$ that is $L_z^+ \approx 282$. Once again, the corresponding low-velocity streak distance from the side-walls show noticeable difference ($d_w^+ \approx 115$ in the closed duct with $A = 2$, and $d_w^+ \approx 96$ in the square closed duct with $Re_b = 4500$).

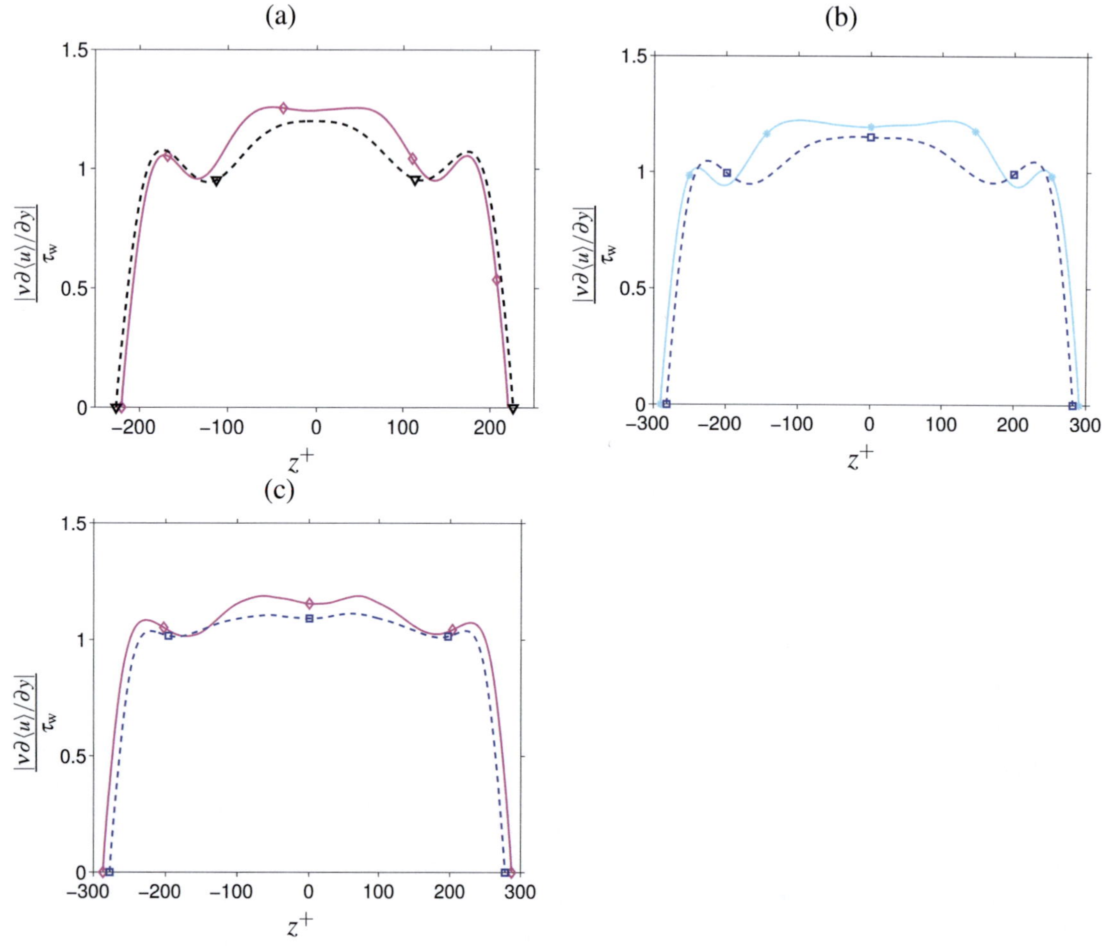

Figure 4.20: Comparisons of mean bottom-wall stress distributions of the duct flow with equivalent friction-based spanwise extension L_z^+, normalised by the whole no-slip wall average value τ_w and presented as a function of z^+. (a) open duct flow with $Re_b = 2205$ and $A = 1.5$ (magenta solid line with ◊), compared with open duct flow with $Re_b = 3500$ and $A = 1$ (black dashed line with ▽). (b) open duct flow with $Re_b = 2205$ and $A = 2$ (cyan solid line with $*$), compared with open duct flow with $Re_b = 4500$ and $A = 1$ (blue dashed line with □). (c) closed duct flow with $Re_b = 2205$ and $A = 2$ (magenta solid line with ◊), compared with closed duct flow with $Re_b = 4500$ and $A = 1$ (blue dashed line with □).

Mean bottom-wall stress distributions of those comparable friction-based spanwise extensions are plotted together in Figure 4.20.

4.4.2 Mean crossflow structure

4.4.2.1 Mean secondary flow

Mean secondary flow patterns of the open duct flow of the current consideration are shown as contour lines of the mean streamfunction $\langle \psi \rangle$ in Figure 4.21. Note that only left-half of the duct cross-sections are shown after the left-right reflective symmetry was applied to the mean field.

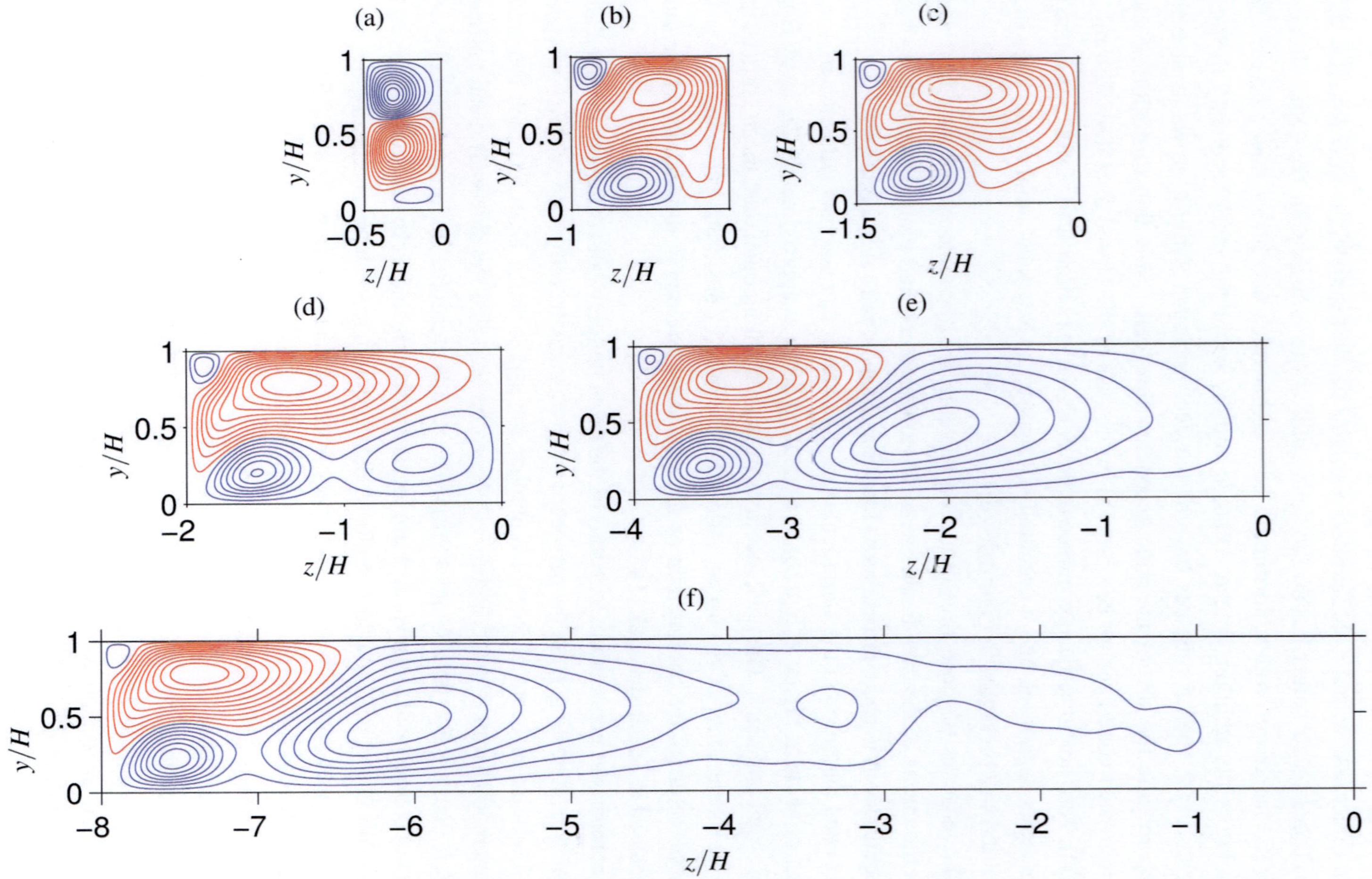

Figure 4.21: Streamlines $\langle\psi\rangle(y,z)$ of **open** duct mean secondary flow computed using $\langle v\rangle$ and $\langle w\rangle$ with reflective symmetry about duct mid-span (with increment $[\max\langle\psi\rangle - \min\langle\psi\rangle]/20$). Red lines correspond to clockwise rotation and blue lines to counter-clockwise motion. (a) $A = 0.5$; (b) $A = 1$; (c) $A = 1.5$; (d) $A = 2$; (e) $A = 4$; (f) $A = 8$, with fixed bulk Reynolds number $Re_b = 2205$. Only left-half of the duct cross-section ($[0,H] \times [-AH,0]$) is shown.

It can be seen that at $A = 0.5$, the open duct mean flow pattern hosts a counter-rotating pair of enlarged mean vortices on the side-wall, together with a small bottom-wall vortex with weak intensity. Subsequently at $A = 1$ as discussed in the previous section, the standard six-vortex pattern, considering the reflective symmetry, appears.
Whilst the open duct flow with $A = 1.5$ exhibits a similar six-vortex state, the flow with $A = 2$, 4, 8 host an additional mean vortex beside the solid-solid corner vortex with the same direction of rotation as the neighbour vortex, forming a new eight-vortex secondary flow pattern. The spanwise location of the additional mean vortex is initially below the outer secondary flow cell in the duct with $A = 2$, then it migrates towards the duct mid-span while it grows its size and becomes comparable to the size of the outer secondary flow cell. Note that formation of such additional bottom-wall vortex cannot be observed in the case of the closed duct flow with $A \geq 2$ (cf. Figure 4.22). Also note that the appearance of the extra mean bottom-wall vortex was found to be Reynolds number dependent, and it eventually disappears as Reynolds number increases (e.g. $Re_b = 5000$ with $A = 2$ presented in Figure 4.9(d)).
Once again, the locations of those streamfunction peaks in the open and closed duct flow were determined in the means of the two-dimensional Newton's method and plotted as distances from the closest boundaries in y- and z-directions as function of aspect ratio A in Figure 4.23.
The aforementioned migration of the additional open duct vortex can be observed between $A = 2$ and 4 (cf. cyan symbols in Figure 4.23(a,b)), then settling down at approximately $2H$ away from the nearest side-wall boundary. The peak locations of the remaining open duct mean vortices are also found to be independent of the aspect ratio after the initial transitional phase up to $A = 2$. Such observation confirms the confinement of the secondary motions in the vicinity of the duct side-walls discussed in Tominaga et al. (1989).
Furthermore, comparison between the locations of the mean bottom-wall vortices in the open and closed duct flow are interesting, from the perspective that they are the only ones exist in both types of duct flow almost independent of Reynolds number and aspect ratio. Note that bulk Reynolds number at around 5000 with $A = 0.5$ and $A = 2$, it can be seen that the side-wall mean vortices start to form (cf. Figure 4.2(c) and Figure 4.9(d)). The results presented in Figure 4.23(a,c) show that the open duct bottom-wall mean vortices are located noticeably closer to the side-walls as well as the bottom-walls in comparison with the closed duct counterparts. More specifically, the open duct vortices are located approximately $0.5H$ away from the nearest side-walls and $0.2H$ above the bottom-wall, whereas the closed duct vortices are approximately $0.85h$ and $0.35h$ away from the side-wall and bottom-wall boundaries respectively.
Moreover, a visual inspection on the mean secondary velocity field of the aforementioned higher Reynolds number open duct flow with $Re_b = 5000$ and $A = 2$ (cf. Figure 4.9(d)) offers a further support to the observation by Tominaga et al. (1989), that the outer secondary flow reaches up to $2H$ in the case of high aspect ratio duct flow. Additional simulations with higher aspect ratios with $Re_b = 5000$ is desirable to determine more precise full spanwise extension of the outer secondary flow since the current mean flow field shows some visible influence from the opposite side-wall in the duct mid-span region below the free-slip plane.

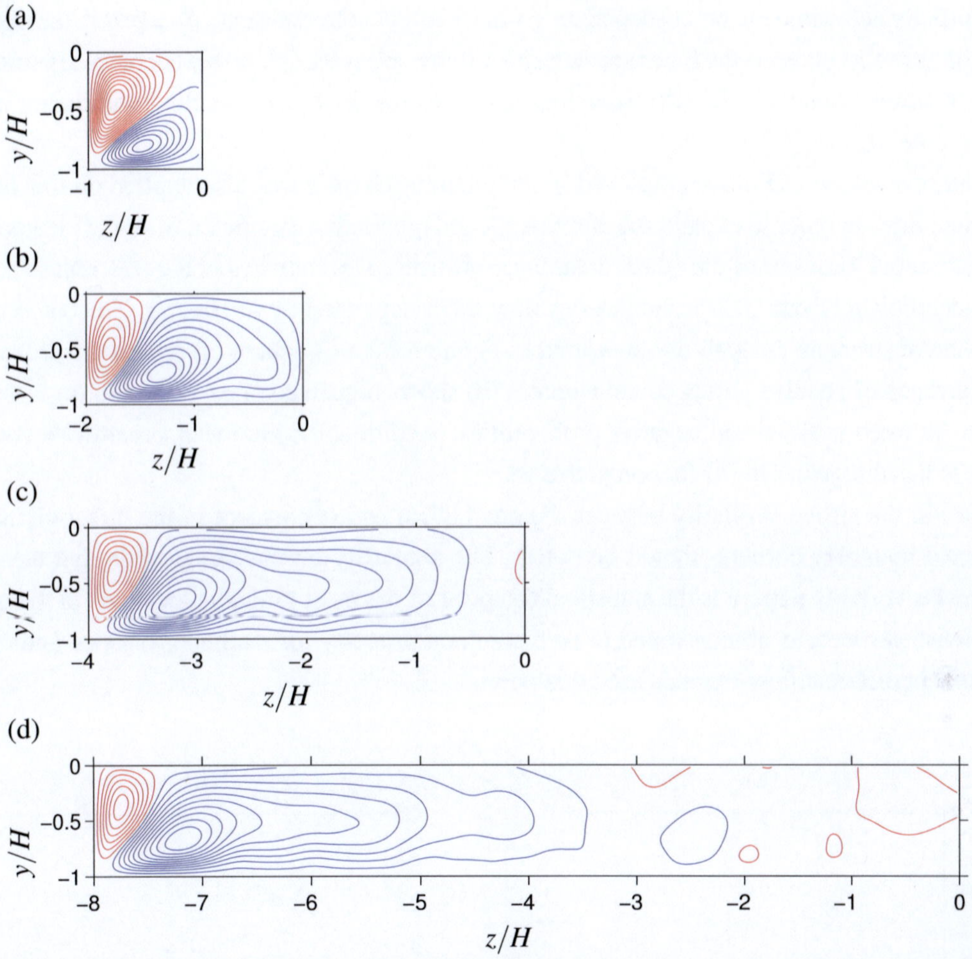

Figure 4.22: Streamlines $\langle\psi\rangle(y,z)$ of **closed** duct mean secondary flow computed using $\langle v\rangle$ and $\langle w\rangle$ with $\pi/2$-rotational symmetry (with increment $[\max\langle\psi\rangle - \min\langle\psi\rangle]/20$). Red lines correspond to clockwise rotation and blue lines to counter-clockwise motion. (a) $A = 1$; (b) $A = 2$; (c) $A = 4$; (d) $A = 8$, with fixed bulk Reynolds number $Re_b = 2205$. Only bottom-left quadrant of the duct cross-section ($[-h,0] \times [-Ah,0]$) is shown.

Similar transitional aspect ratio dependence on the secondary flow cell peak locations can be observed in case of the closed duct flow (cf. Figure 4.23(c,d)), although the moderate aspect ratio dependence still exists even in the widest closed duct with $A = 8$. Additional direct numerical simulations of closed duct flow with larger aspect ratios are therefore required to conclude this part of the investigation.

4.4.2.2 Mean streamwise vorticity

Corresponding mean streamwise vorticity contours are presented in Figure 4.24 for the open duct flow, and in Figure 4.25 for the closed duct flow. Consequently, the peak locations of those profiles are summarised as function of aspect ratio in Figure 4.26. Analogous to the trend observed in the previous investigations on the secondary flow streamfunction profiles and their peak locations, the mean streamwise vorticity profiles maintain a similarity to the ones observed in the reference open duct case with $A = 1$. Only exception is again in the vorticity profile of the duct with $A = 0.5$ (Figure 4.24(a)), where the inner secondary vorticity cell forms a equally-sized pair with the side-wall vorticity cell while the outer secondary vorticity cell becomes absent. As a result, the distance from the free-slip plane to the inner secondary vorticity cell peak ($s^+ \approx 40$) becomes significantly larger in comparison with the corresponding distances for the open duct flow with larger aspect ratios ($s^+ \approx 10$).

The vortex eduction technique employed in the previous chapter was also applied on this narrow open duct flow, in order to explain the aforementioned qualitative difference from the viewpoint of the preferential locations of the quasi-streamwise vortices. The statistics of the coherent structures were accumulated from 1000 instantaneous flow fields separated by $\approx 10\nu/u_\tau^2$. The outcomes of the coherent structure analysis are presented in Figure 4.27(a-c), where (a) represents probability of occurrence of positive vortex cores; whereas (b) shows negative vortex core; and (c) is the difference between positive and negative probabilities. Additionally, the mean streamwise vorticity profile is listed together in (d) for completeness.

First of all, the strong similarity between Figure 4.27(c) and (d), except in the direct vicinity of the mixed-boundary corners, should be noted. The similarity implies once again that the mean streamwise vorticity pattern is the statistical footprint of the most probable locations of the quasi-streamwise vortices, as demonstrated to be a common property for a number of open and closed duct flow in different flow regimes until this point.

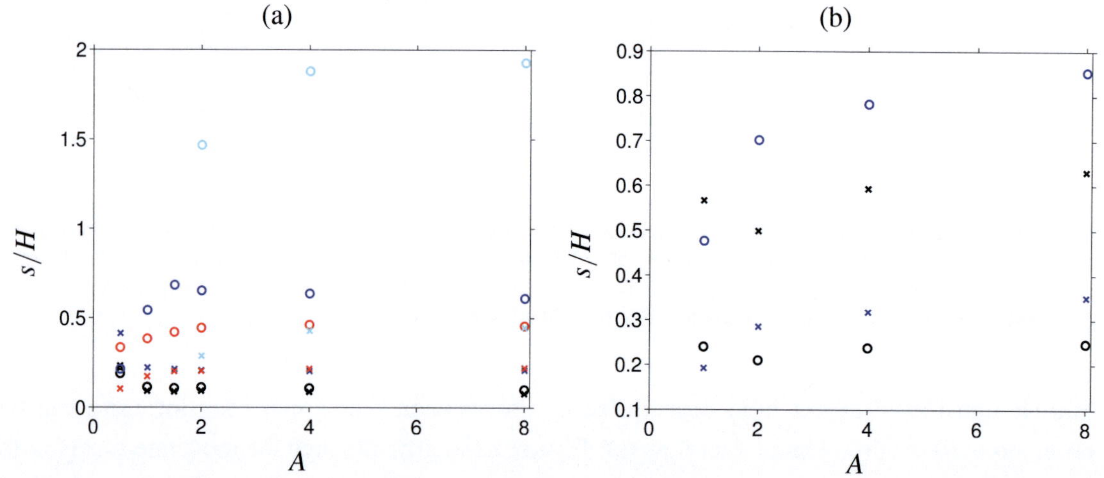

Figure 4.23: Positions of extrema of $\langle\psi\rangle(y,z)$ in external coordinates as function of A. (a) open duct; (b) closed duct with bulk Reynolds number $Re_b = 2205$. Colours in (a) indicate: black, inner secondary flow cell; blue, outer secondary flow cell; red, bottom-wall cell; cyan, second bottom-wall cell. Symbols: $\circ$ distance to $z/H = -A$; $\times$ distance to $y/H = 1$ (blue symbols) and $y/H = 0$ (red and cyan symbols). Colours in (b) indicate: black, side-wall cell; blue, bottom-wall cell. Symbols: $\circ$ distance to $z/h = -A$; $\times$ distance to $y/h = -1$.

In contrast to the similarity with the other duct flow configurations, however, the relatively high probability peaks of the occurrence of the vortices rotating in the opposite direction to the rotational direction of the mean inner secondary flow are unique phenomenon to this narrow open duct configuration (cf. the peak in the top-left corner of Figure 4.27(a), and in the top-right corner of (b)). In combination with the extended probability "spikes" of the dominant inner secondary peaks penetrating deep into the mixed-boundary corners, the difference in the probability presented in Figure 4.27(c) shows very distinctive alternating pattern of the vortex probability peaks in the mixed-boundary corners that has not been observed elsewhere. Note, however, that such alternating patterns in the mixed-boundary corners are not reflected in the mean streamwise vorticity statistics (cf. Figure 4.27(d)), where only the statistical signatures of the spikes-shaped probabilistic distribution are visible. Further investigations are required to determine the source of the discrepancy, for example by taking into account the intensity of the educed coherent structures that are neglected in the current technique.
The probability peaks corresponding to the inner secondary vorticity cells mark the highest values as in the case of $A = 1$, although the difference from the other probability peaks is not as extreme as in the reference case. The former observation implies that the interaction between quasi-streamwise vortices and the free-slip plane still plays an important role to sort and contain the vortices with a certain sense of rotation in the mixed-boundary corners in this narrow duct configuration. The latter observation, on the other hand, implies that the strict geometrical constraint in spanwise direction imposed by the side-wall boundaries in the case of the narrow duct configuration restricts the movements and the locations of quasi-streamwise vortices as strong as the above free-slip plane interaction. In other words, the degree of the locational preference of the quasi-streamwise vortices in the mixed-boundary corner regions is relatively less dominant, being comparable to the one in the other flow regions. From the above observation, it is possible to conjecture that the enlarged inner secondary vorticity cell in the current narrow duct configuration is a consequence of the probability of the existence of the mixed-boundary corner vortices being more distributed, resulting the mixed-boundary corner streamwise vorticity peaks to be less concentrated in both intensity and cross-sectional area.
To this end, it is important to recall that the current vortex eduction technique only takes into account the location of the detected vortex cores, therefore any conjectures related to the size of those structures cannot be confirmed at this point. For example in the above conjecture, we implicitly assumed that the near-wall vortices in the mixed-boundary corner regions are not significantly larger than the standard near-wall vortex dimensions, although the size of the resulted inner secondary vorticity cells is noticeably larger. Even though this assumption can be justified by the fact that the scale-separation in the current configuration should be moderate due to the low Reynolds number, it is still necessary in our future work to confirm such assumption by employing a different kind of vortex eduction technique with the capability of taking into account the cross-sectional area of those vortices.

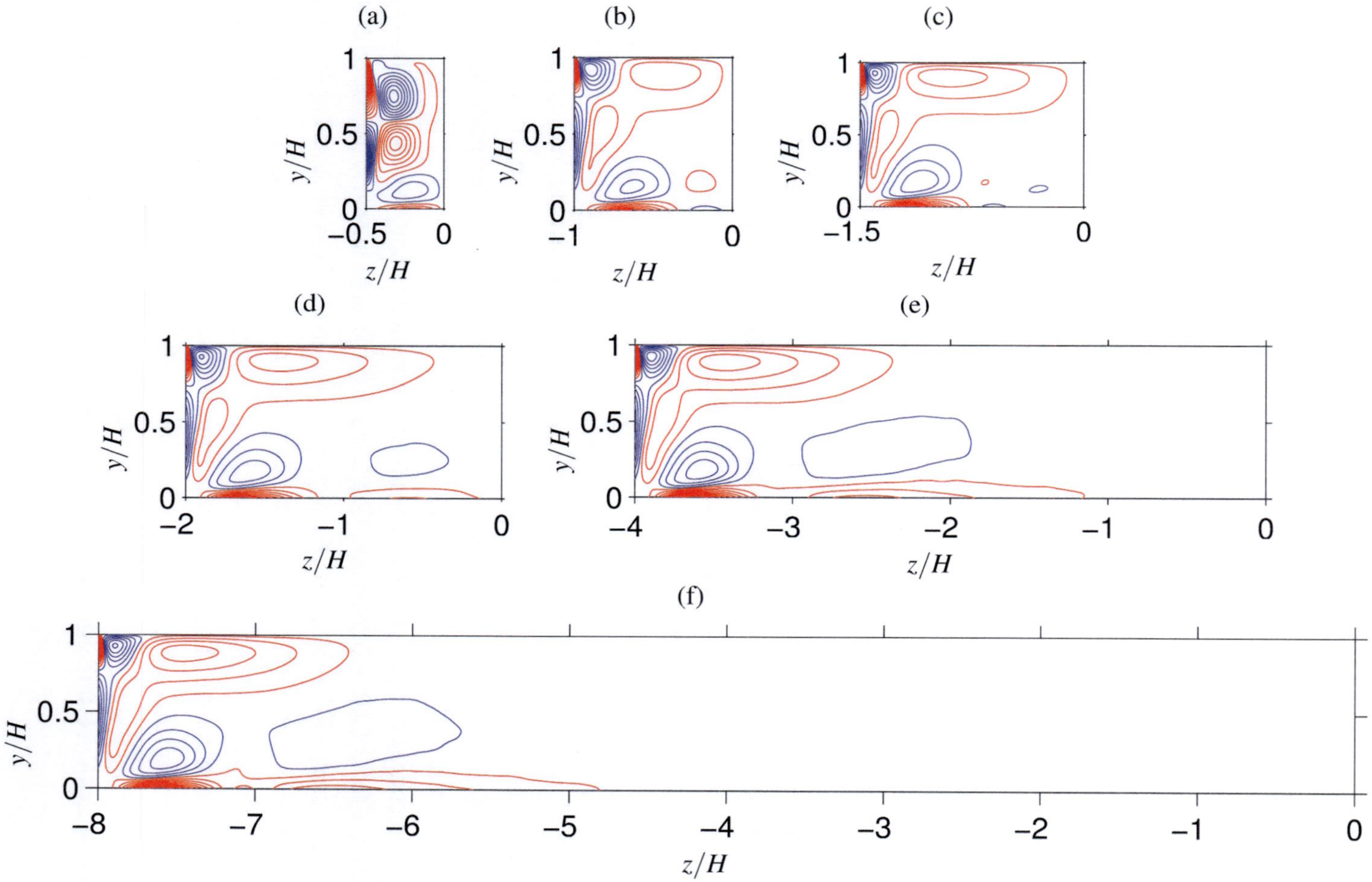

Figure 4.24: Contours of $\langle \omega_x \rangle (y,z)$ of **open** duct mean secondary flow computed using $\langle v \rangle$ and $\langle w \rangle$ with reflective symmetry about duct mid-span (with increment $[\max\langle \omega_x \rangle - \min\langle \omega_x \rangle]/20$). Red lines correspond to positive values and blue lines to negative ones. (a) $A = 0.5$; (b) $A = 1$; (c) $A = 1.5$; (d) $A = 2$; (e) $A = 4$; (f) $A = 8$, with fixed bulk Reynolds number $Re_b = 2205$. Only left-half of the duct cross-section ($[0,H] \times [-AH,0]$) is shown.

4.4.2.3 Secondary flow intensity and its decay property

In order to quantify the confinement of the open duct secondary motions in the vicinity of the duct side-walls, we first investigate the spanwise distribution of the cross-streamwise velocity scale, deduced from the kinetic energy of the mean secondary motions, viz.

$$E_{vw}(z) = \frac{1}{2H}\int_0^H \left(\langle v\rangle^2 + \langle w\rangle^2\right) dy \tag{4.6}$$

Additionally, the corresponding quantity of the closed duct flow will be investigated for comparison, where the lower and upper limits of of the integral in y-direction are replaced respectively by $-h$ and $+h$.

The secondary flow velocity distributions are normalised by bulk velocity u_b, and presented as function of distance from the duct side-walls ($z'' = AH - |z|$) in Figure 4.28, for (a) the open and (b) closed duct flow with $Re_b = 2205$.

It turns out that the secondary velocity intensity decays exponentially with the distance from the side-walls increases (i.e. $\sqrt{E_{vw}} \sim \exp(-\lambda z'')$), and the decay constant λ is considered to be asymptotically approaching towards some value, such as $\lambda \approx 0.5$ for the current open duct flow with $A = 8$. A consequence of the increased Reynolds number remains unknown at this time of writing.

Moreover, the decay constant was found to be higher in the closed duct flow, for instance, the closed duct with the largest aspect ratio of $A = 8$ exhibits $\lambda \approx 0.7$. A structural explanation for the slower decay of the secondary velocity in the open ducts can be offered from the fact that the outer secondary flow cells, which are unique to the open duct flow, are shifted and extended towards the duct mid-span far beyond where the closed duct side-wall vortices —which are equivalent to the outer secondary flow cells in a sense of the rotational direction— have their reach. Also, it is worth recalling at this point the earlier observation in the previous chapter that the maximum open duct secondary flow velocity is consistently found on the free-slip plane, whereas the equivalent locations in the closed duct flow are always found around the solid-solid corner diagonals. As a result, the spanwise peak locations in the open duct flow are shifted towards the duct mid-span with a noticeable distance (e.g. the maximum peak locations of $\sqrt{E_{vw}}$ presented in Figure 4.29 are at $\approx 0.54H$ and $\approx 0.42H$ in the open and closed duct flow respectively). Additionally, the magnitudes of the maximum secondary flow velocities in the two configurations differ significantly: the open duct maximum secondary flow intensity is almost as twice as the intensity in the closed duct flow (cf. Figure 4.30). Consequently, the kinetic energy of secondary motion is shifted further in spanwise direction in the open duct flow.

Next, let us investigate the mean secondary velocity that is integrated not only in y-direction, but also some distance from the side-wall boundary in z-direction. We shall define side-wall cross-plane area average mean secondary intensity $U_\perp^*$ as follows:

$$U_\perp^* = \left[\frac{1}{2aH^2}\int_{-aH}^{+aH}\int_0^H \left(\langle v\rangle^2 + \langle w\rangle^2\right) dy dz\right]^{1/2}, \tag{4.7}$$

and alternatively for closed ducts:

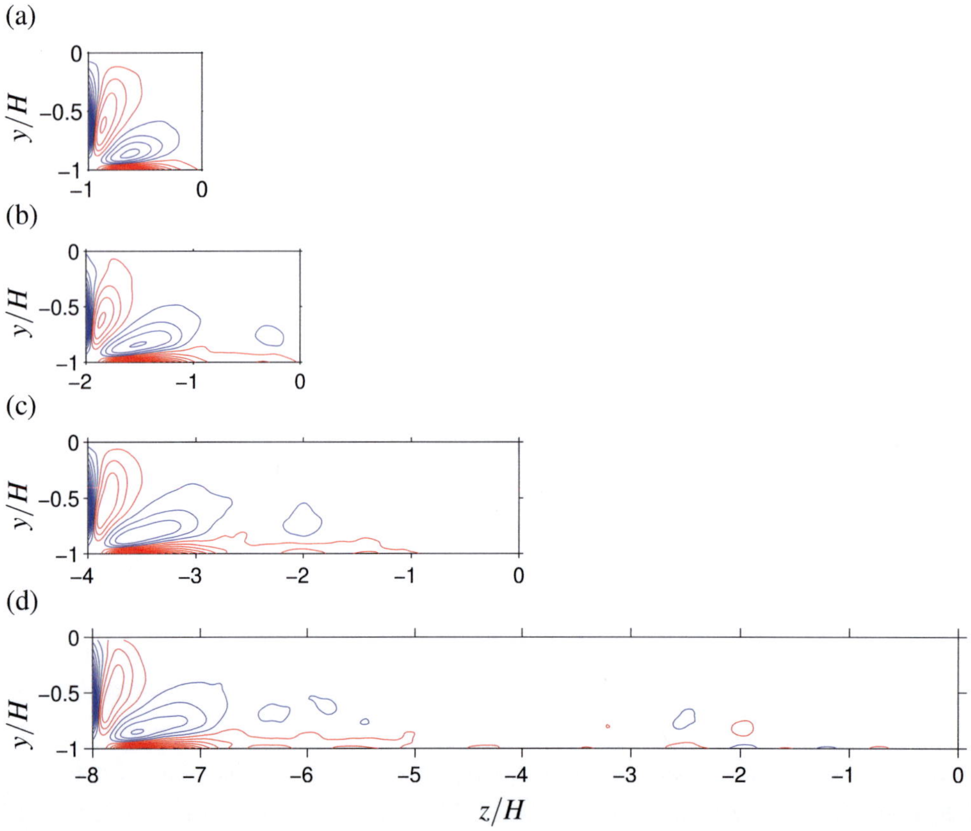

Figure 4.25: Contours of $\langle\omega_x\rangle(y,z)$ of **closed** duct mean secondary flow computed using $\langle v\rangle$ and $\langle w\rangle$ with $\pi/2$-rotational symmetry (with increment $[\max\langle\omega_x\rangle - \min\langle\omega_x\rangle]/20$). Red lines correspond to positive values and blue lines to negative ones. (a) $A=1$; (b) $A=2$; (c) $A=4$; (d) $A=8$, with fixed bulk Reynolds number $Re_b = 2205$. Only bottom-left quadrant of the duct cross-section ($[-h,0]\times[-Ah,0]$) is shown.

$$U_\perp^* = \left[\frac{1}{4ah^2}\int_{-ah}^{+ah}\int_{-h}^{+h}(\langle v\rangle^2 + \langle w\rangle^2)dydz\right]^{1/2}, \tag{4.8}$$

where a is a given side-wall distance coefficient $0 < a \leq A$. Please note that the original definition of the above side-wall averaged secondary flow intensity was first proposed by Wakabayashi for closed duct flow (*personal communication with Prof. Kawahara*).

Figure 4.31 and 4.32 show the value of $U_\perp^*$ as function of the value of a on the open and closed ducts respectively. It is important to notice that the peak values of $U_\perp^*$ occur with $a \leq 1$ (cf. Figure 4.33 for the value of a corresponding to the maximum $U_\perp^*$ for each A), which again confirms quantitatively that the secondary motions are confined within the relatively narrow side-wall regions (i.e. $z'' \sim H(h)$).

Finally, let us fix the value of a at unity and investigate the side-wall averaged secondary intensity for the current range of aspect ratios. Figure 4.34 shows $U_\perp^*$ normalised by the bulk streamwise velocity as function of aspect ratio for: (a) the open and (b) closed ducts, plotted alongside the secondary flow intensity integrated over the whole duct cross-section (i.e. $U_\perp/u_b$, black dashed lines). It can be seen that the near side-wall secondary flow intensities can be approximated as a constant independent of the aspect ratio in both open and closed duct flow.

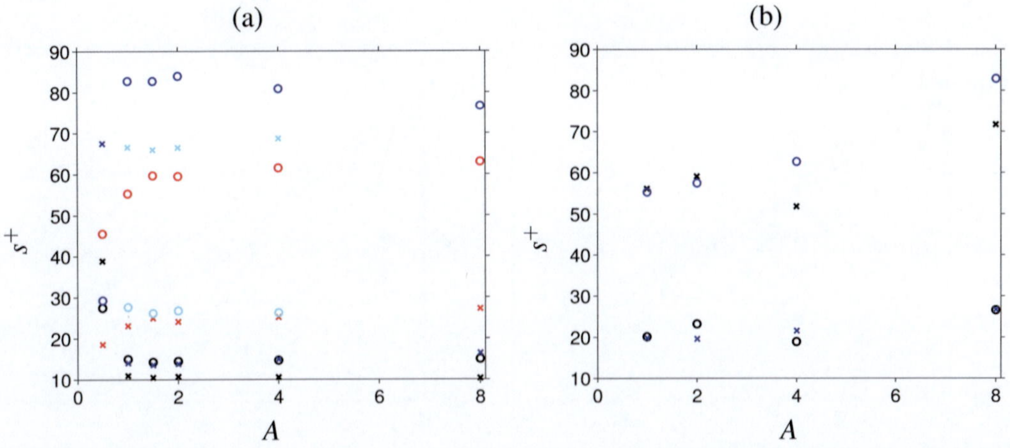

Figure 4.26: Positions of extrema of $\langle\omega_x\rangle(y,z)$ in wall coordinates as function of A. (a) open duct; (b) closed duct with bulk Reynolds number $Re_b = 2205$. Colours in (a) indicate: black, inner secondary vorticity cell; blue, outer secondary vorticity cell; red, bottom-wall vorticity cell; cyan, side-wall vorticity cell. Symbols: $\circ$ distance to $z/H = -A$; $\times$ distance to $y/H = 1$ (blue symbols) and $y/H = 0$ (red and cyan symbols). Colours in (b) indicate: black, side-wall vorticity cell; blue, bottom-wall vorticity cell. Symbols: $\circ$ distance to $z/h = -A$; $\times$ distance to $y/h = -1$.

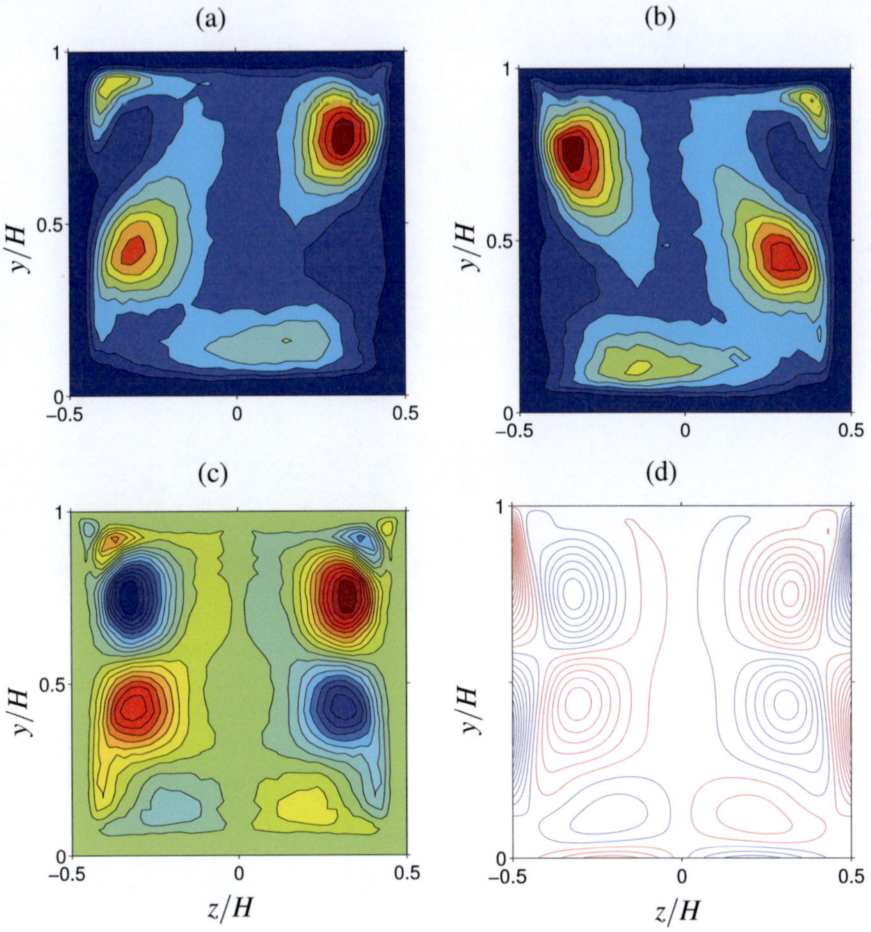

Figure 4.27: (a-c) Probability of occurrence of vortex centres for the open duct case with $Re_b = 2205$ and $A = 0.5$, detected by the technique proposed by Kida and Miura (1998). (a) vortices with *positive* streamwise vorticity; (b) *negative* streamwise vorticity; (c) difference of (a) and (b), being symmetrised about the duct mid-span. (d) Iso-contours of $\langle\omega_x\rangle(y,z)$ (replotted for convenience). The iso-contours indicate: (a,b) 0.1(0.1)0.9 times the maximum values; (c) -0.9(0.1)0.9 times the maximum values; (d,e) 0.1(0.05)0.9 times the maximum values respectively.

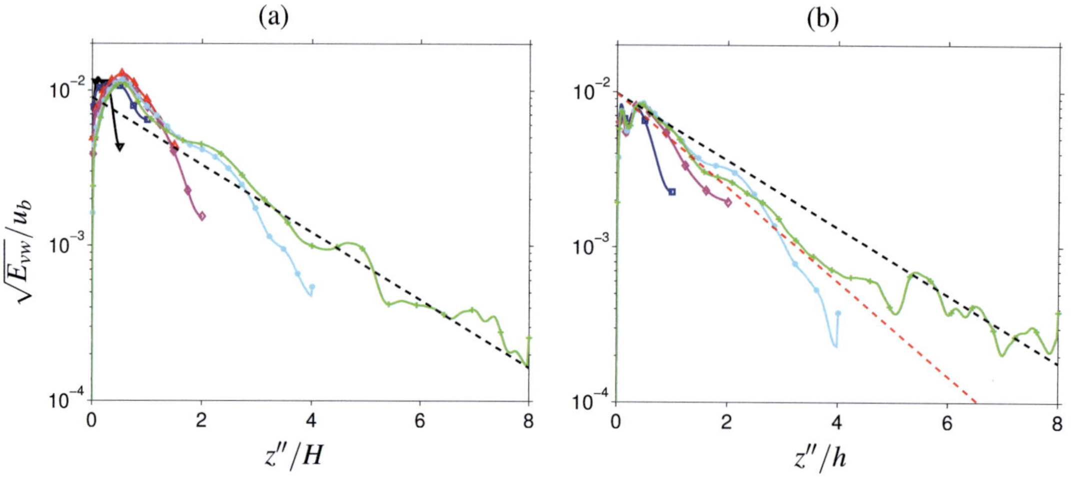

Figure 4.28: Spanwise distribution of secondary flow energy $E_{vw} = \frac{1}{2}\int_0^H (\langle v\rangle^2 + \langle w\rangle^2)\, dy$ as a function of distance from side-wall $z''/H = (AH - |z|)/H$, of (a) open and (b) closed duct flow. Reynolds number is fixed at $Re_b = 2205$ and coloured symbol code is as of Figure 4.19. Black dashed lines indicate $\approx \exp(-0.5z''/H)$, whereas red line in (b) indicates $\approx \exp(-0.7z''/h)$.

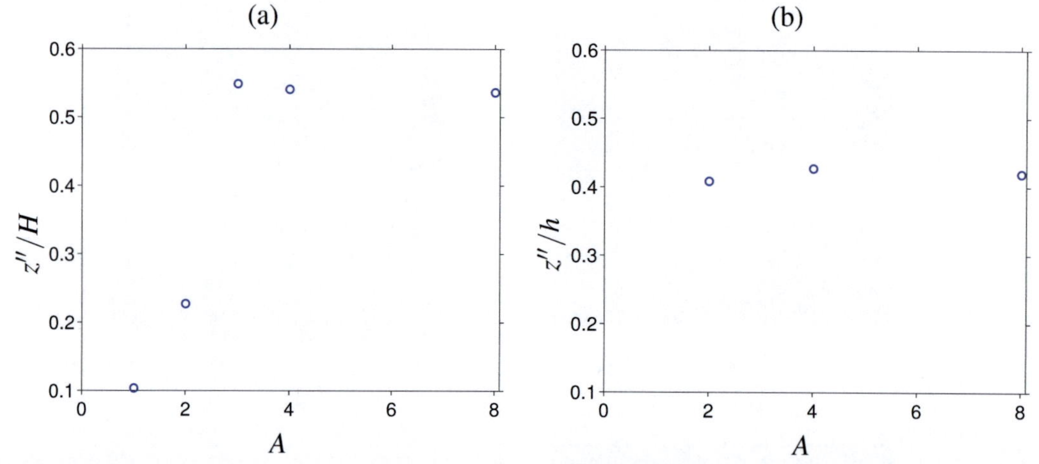

Figure 4.29: Side-wall distance of the maximum of $\sqrt{E_{vw}}$ as function of aspect ratio A. (a) open and (c) closed duct flow with $Re_b = 2205$.

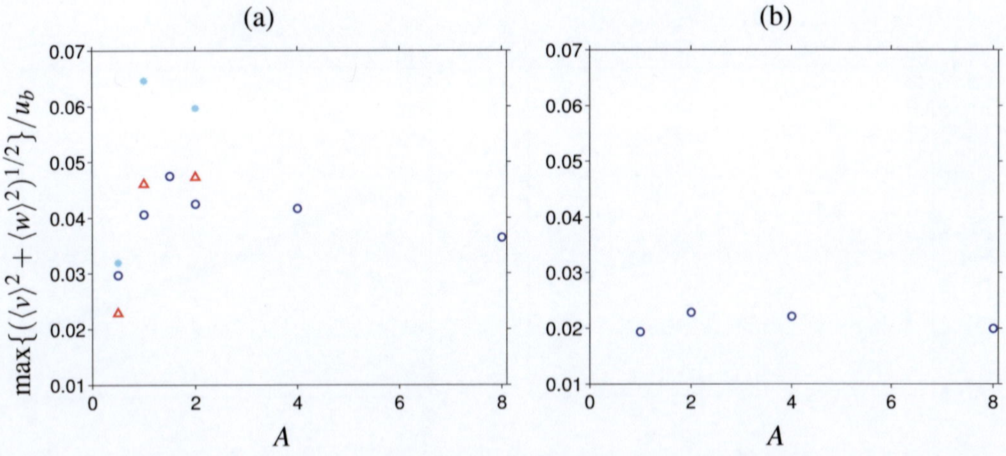

Figure 4.30: Intensities of maximum mean secondary velocity $\max\{(\langle v\rangle^2+\langle w\rangle^2)^{1/2}\}$ as function of aspect ratio A, of (a) open and (b) closed duct flow. Coloured symbols indicate different bulk Reynolds number Re_b: ○, $Re_b = 2205$; △, $Re_b = 3000$; $*$, $Re_b = 5000$.

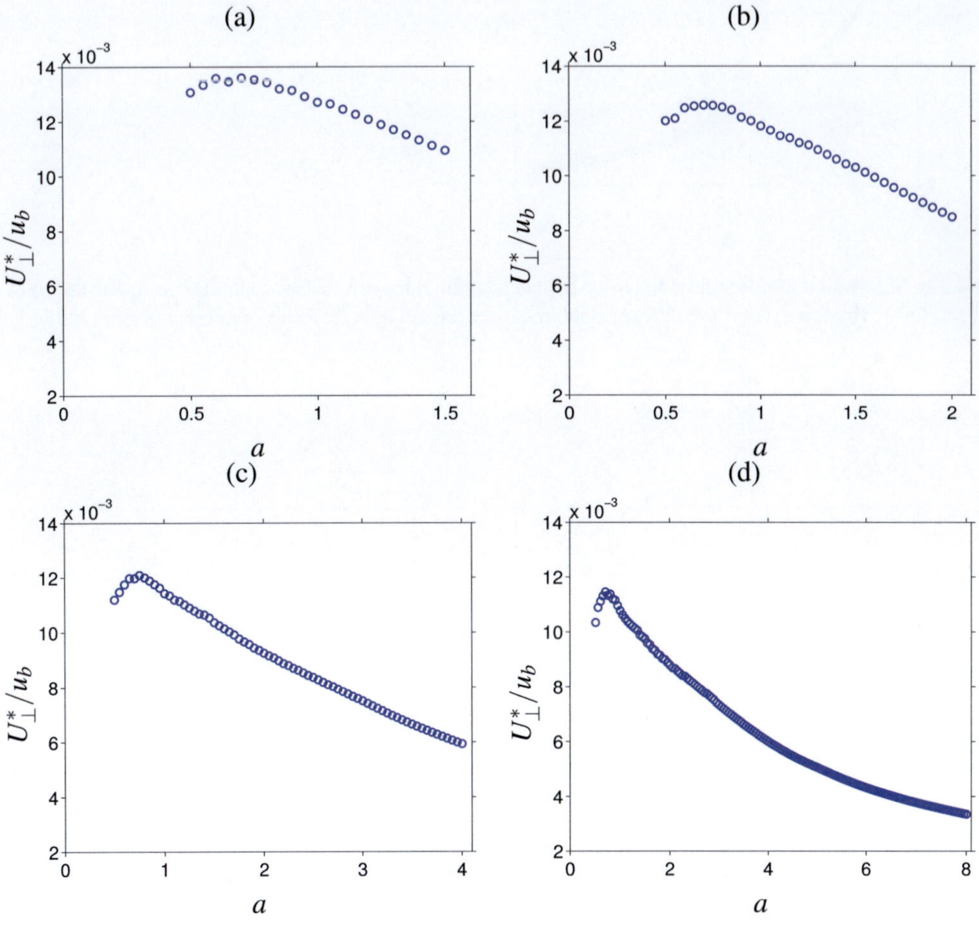

Figure 4.31: Side-wall secondary flow intensity dependence to side-wall coefficient a, for different aspect ratio A: (a) $A = 1.5$; (b) $A = 2$; (c) $A = 4$; (d) $A = 8$. Open duct flow with a fixed bulk Reynolds number at $Re_b = 2205$.

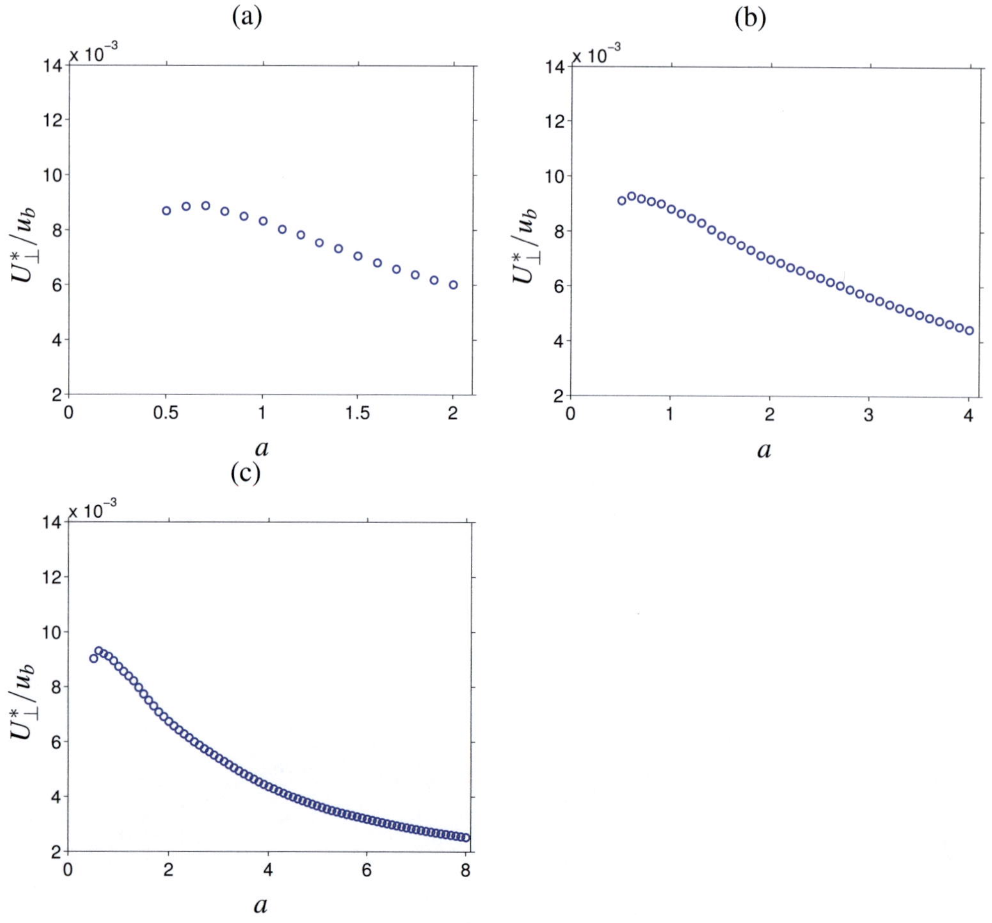

Figure 4.32: Side-wall secondary flow intensity dependence to side-wall distance coefficient a, for different aspect ratio A: (a) $A = 2$; (b) $A = 4$; (c) $A = 8$. Closed duct flow with a fixed bulk Reynolds number at $Re_b = 2205$.

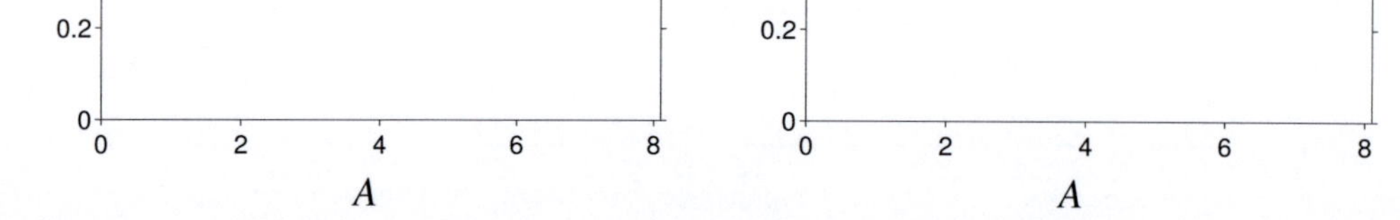

Figure 4.33: Side-wall distance coefficient a of the maximum of $U_\perp^*$ as function of aspect ratio A. (a) open and (b) closed duct flow with $Re_b = 2205$.

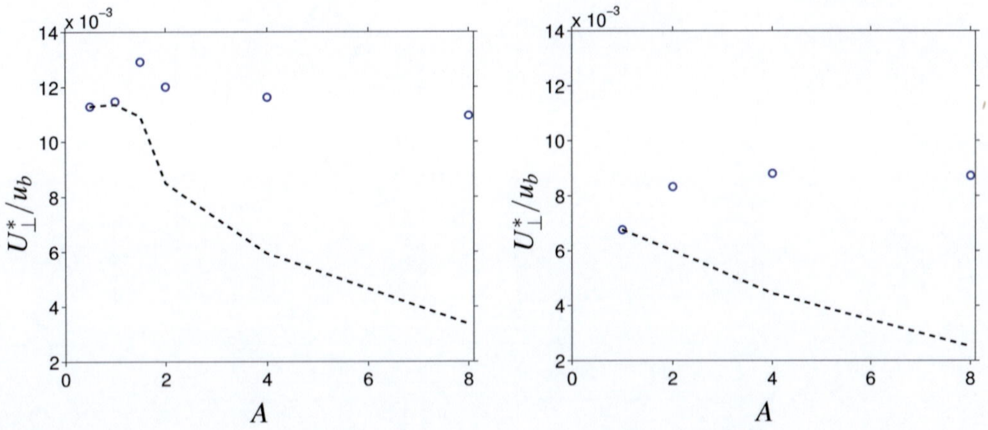

Figure 4.34: Side-wall secondary flow intensity $U_\perp^*$ vs. aspect ratio A. Black dashed line indicates the corresponding full secondary flow intensities $U_\perp$. Value of a is fixed at unity, whereas Re_b is fixed at 2205. (a) open, (b) closed duct flow.

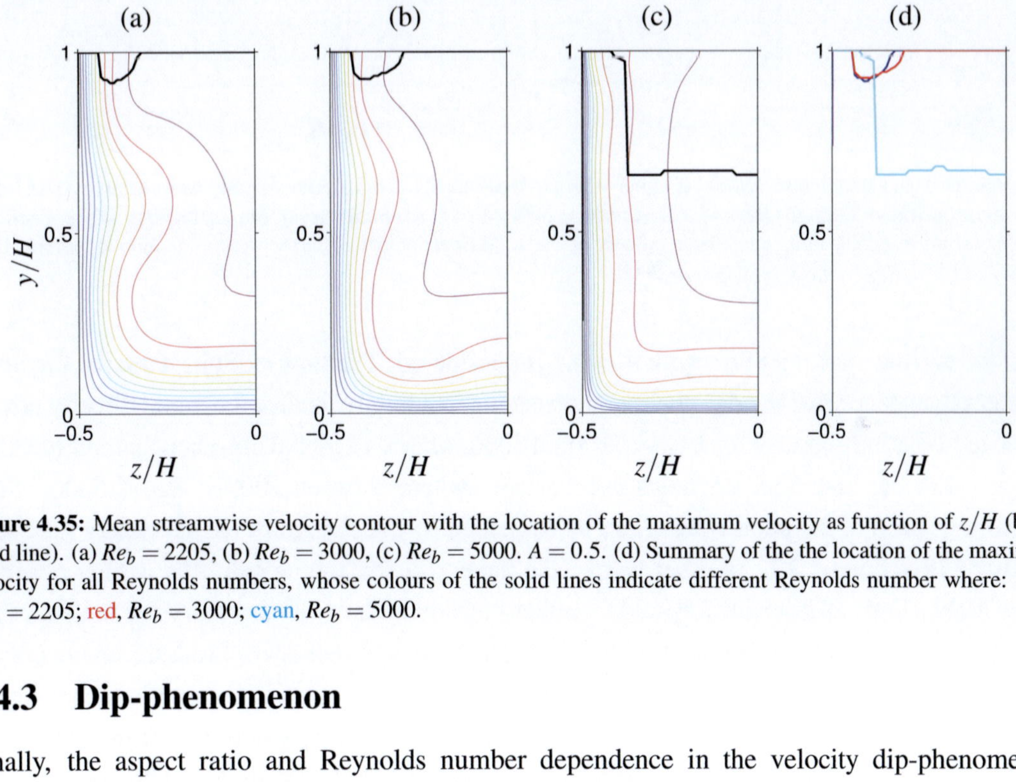

Figure 4.35: Mean streamwise velocity contour with the location of the maximum velocity as function of z/H (black solid line). (a) $Re_b = 2205$, (b) $Re_b = 3000$, (c) $Re_b = 5000$. $A = 0.5$. (d) Summary of the the location of the maximum velocity for all Reynolds numbers, whose colours of the solid lines indicate different Reynolds number where: blue, $Re_b = 2205$; red, $Re_b = 3000$; cyan, $Re_b = 5000$.

4.4.3 Dip-phenomenon

Finally, the aspect ratio and Reynolds number dependence in the velocity dip-phenomenon will be discussed. Here we distinguish the "*global dip-phenomenon*" from the "*localised dip-phenomenon*", where the former refers to the submergence of the location of the mean streamwise velocity maxima even at the duct mid-span ($z/H = 0$), whereas the latter refers to the localised phenomenon observed only close to the duct side-walls. Figure 4.35, 4.36 and 4.37 visualise the spanwise distribution of the locations of the mean streamwise velocity maxima in y-direction (indicated by black solid lines), for the three different aspect ratios ($A = 0.5$, 1 and 2), each simulated with the three different bulk Reynolds numbers ($Re_b = 2205$, 3000 and 5000) respectively. Contour lines of the mean streamwise velocities are plotted together for completeness.

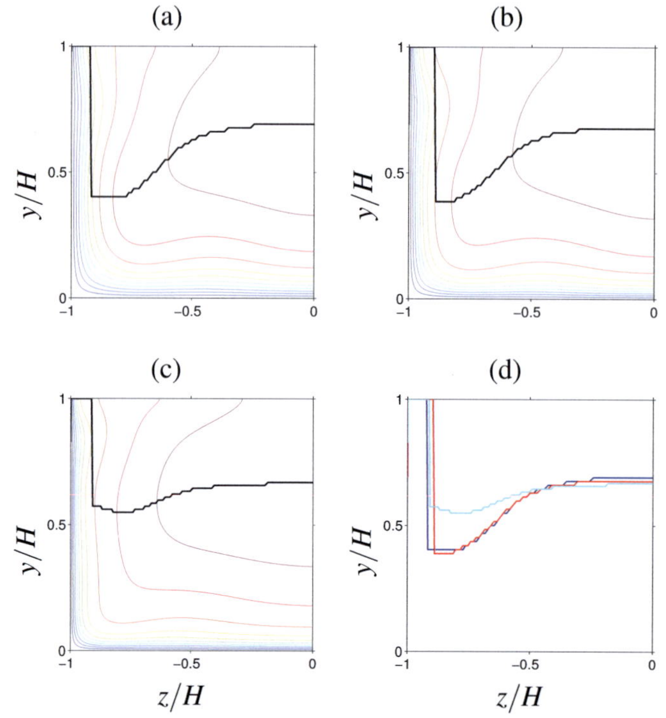

Figure 4.36: Mean streamwise velocity contour with the location of the maximum velocity as function of z/H (black solid line). (a) $Re_b = 2205$, (b) $Re_b = 3000$, (c) $Re_b = 5000$. $A = 1$. (d) Summary of the the location of the maximum velocity for all Reynolds numbers, whose colours of the solid lines indicate different Reynolds number where: blue, $Re_b = 2205$; red, $Re_b = 3000$; cyan, $Re_b = 5000$.

First, the narrow open duct flow with $A = 0.5$ is considered. The flow exhibits a unique Reynolds number dependence that the dip-phenomenon occurs in a highly localised manner directly next to the mixed-boundary corners for $Re_b = 2205$ and 3000, whereas a global dip-phenomenon suddenly appears with the flow bulk Reynolds number somewhere between $3000 < Re_b \leq 5000$. Such radical transition of the phenomenon can be explained by the change of the secondary flow state discussed in Section 4.3.1. In other words, the outer secondary flow cell, that is only observed at the higher limit of the bulk Reynolds number considered here, is necessary for the duct flow with $A = 0.5$ to achieve the global dip-phenomenon at the duct mid-span. From the context of the preferential locations of the large-scale low-velocity streaks discussed in the previous chapter, the existence of the mean outer secondary flow is a necessary condition to transport the large-scale streaks from the side-wall regions towards the duct core regions below the free-slip plane.

As discussed in the previous chapter, all open duct flow with $A = 1$ achieve the global dip-phenomenon except for the minimal Reynolds number at $Re_b = 1450$. The reason for the absence of the global dip-phenomenon in the minimal Re flow was explained from the viewpoint of the preferential locations of the large-scale low-velocity scales around the duct mid-span directly below the free-slip plane (cf. Section 3.3.5.2). When we consider the corresponding spanwise velocity dip profiles, a sharp initial drop near the side-wall regions exists with a visible Reynolds number dependence up to $z'' \approx 0.5H$, followed by a plateau region which continues to the duct mid-span, whose profile is Reynolds number independent once recovered from the initial drop

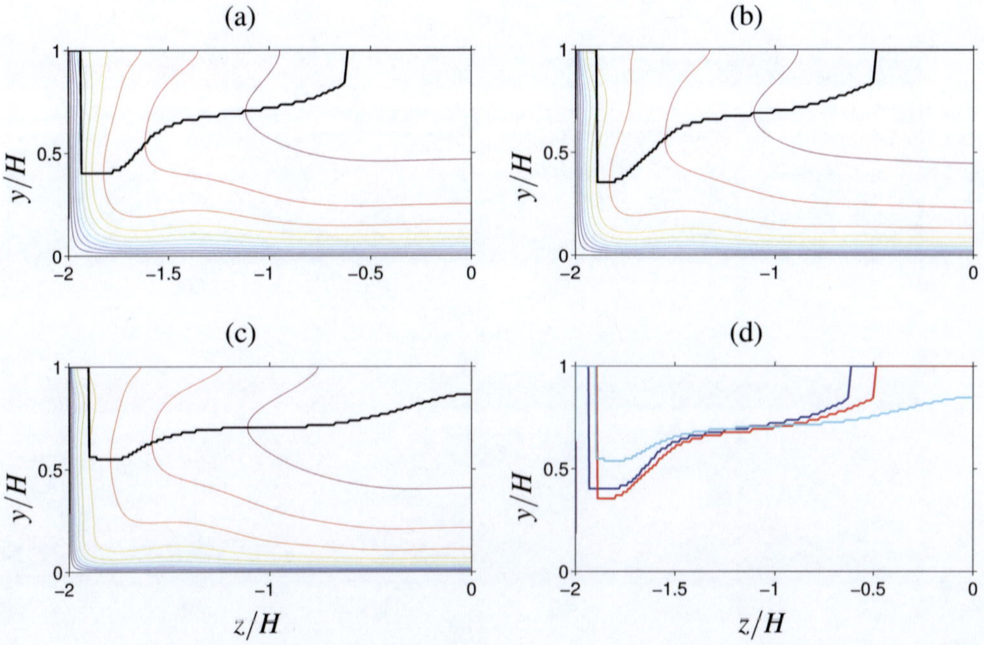

Figure 4.37: Mean streamwise velocity contour with the location of the maximum velocity as function of z/H (black solid line). (a) $Re_b = 2205$, (b) $Re_b = 3000$, (c) $Re_b = 5000$. $A = 2$. (d) Summary of the the location of the maximum velocity for all Reynolds numbers, whose colours of the solid lines indicate different Reynolds number where: blue, $Re_b = 2205$; red, $Re_b = 3000$; cyan, $Re_b = 5000$.

phase. More specifically on the Re-dependence in the initial drop, the depth is considerably shallower for the $Re_b = 5000$ case than the other two cases with $Re_b = 2205$ and 3000.

The duct flow with $A = 2$ shows a similar picture with the initial drop near the side-walls, where the highest Reynolds number flow ($Re_b = 5000$) experiences considerably shallower initial drops compared to the other two lower Reynolds number cases. Subsequently, the Reynolds number independent plateau region follows as before, extending up to $z'' \approx H$. Beyond that point, the location of dip-phenomenon starts to rise again, gradually towards the free-slip plane for all three Reynolds numbers, with some visible Reynolds number dependence in the angles of the rising steeper for lower Reynolds numbers. In the following region close to the duct mid-span, the velocity dip-phenomenon differs even qualitatively depending on the flow Reynolds number. For the flow with $Re_b = 2205$ and 3000, the velocity dip-phenomenon suddenly disappears around $z''/H \approx 1.4$ and ≈ 1.5 respectively. Once the flow Reynolds number is further increased up to $Re_b = 5000$, the gradual rise of the dip location continues until the duct mid-span, and a global dip-phenomenon is achieved as a result.

The observed results imply the following points: (i) the spanwise locations where the dip-phenomenon occurs extend from the duct side-walls; (ii) the side-wall distance where the dip-phenomenon is in effect is proportional to Reynolds number, at least for the moderate range of Reynolds numbers considered here; (iii) the critical aspect ratio for the global dip-phenomenon is also function of Reynolds number, at least for the considered Reynolds number range.

Note that the point (i) had been widely known prior to this work, and served as a foundation of the dip-phenomenon models that were discussed in the previous chapter (cf. Equation 3.15 for the empirical model by Wang et al. (2001), and Equation 3.16 for the model by Yang et al.

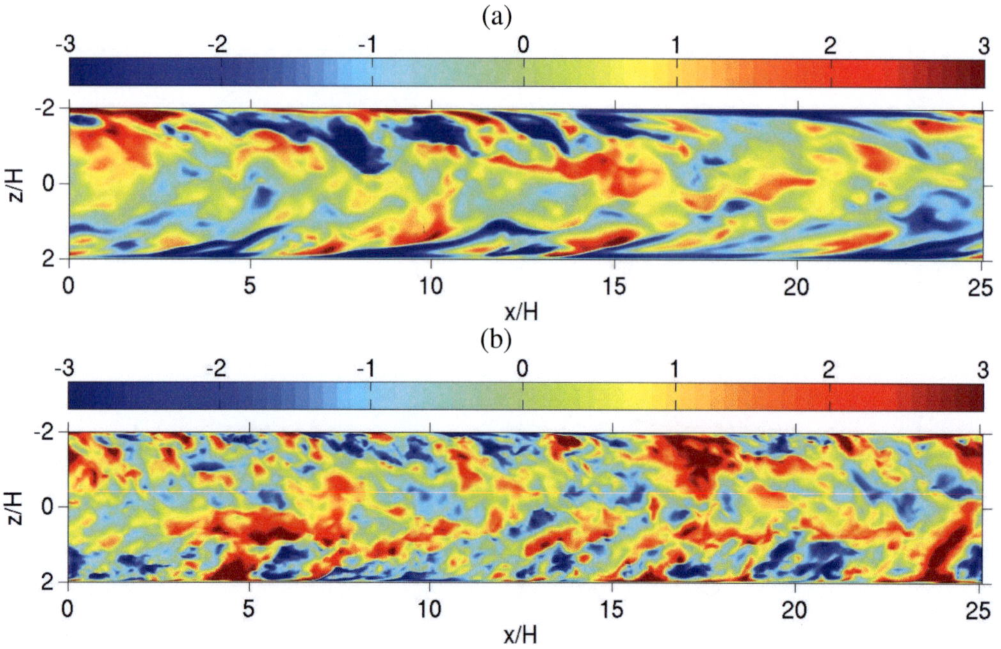

Figure 4.38: Instantaneous streamwise velocity fluctuation field normalised by friction velocity (u'/u_τ), plotted on xz-plane approximately $10\delta_v$ below the free-slip plane, in open duct flow with $A = 2$. (a) $Re_b = 2205$, (b) $Re_b = 5000$.

(2004)). A structural explanation discussed for the phenomenon in the previous chapter, as a consequence of low-velocity streaks from the side-walls transported below the free-slip plane towards the duct mid-span (cf. Section 3.3.5.2), also coincides with the side-wall origination of the dip-phenomenon.

On the other hand, the Reynolds number dependence in the point (ii), and (iii) as its consequence, have not been reported elsewhere to our knowledge, most certainly due to that almost all other previous studies on this topic considered considerably higher Reynolds numbers than the current simulations. Whilst our preliminary study indicates a link between the flow Reynolds number and the extent of the oblique bands of low-velocity streaks from the side-walls observed only in the direct vicinity of the free-slip plane (cf. Figure 4.38), more detailed structural investigations are required for a better understanding of the reported phenomenon.

The point (iii) also implies premature disappearances of the global dip-phenomenon with respect to the critical aspect ratio reported in Nezu and Rodi (1985) and Tominaga et al. (1989) at $A_{\text{crit}} \approx 2.5$, and also the value predicted by the model by Wang et al. (2001) at 2.63 (cf. Equation 3.15).

The observed Reynolds number dependence in the submerged depths of the global dip-phenomenon are summarised in Figure 4.39 for all considered aspect ratios. As a reference, the representative critical aspect ratio at $A_{\text{crit}} = 2.5$ is also plotted for comparison. As it was discussed in the previous chapter, some Reynolds number dependence in the submerged depth can be observed at $A = 1$. However, since the flow with other aspect ratio did not exhibit the global dip-phenomenon for more than one Reynolds number, the submerged depth scaling of those aspect ratio configurations cannot be determined in the current study. It is therefore desirable to perform higher Reynolds number simulations for those aspect ratios in the future studies.

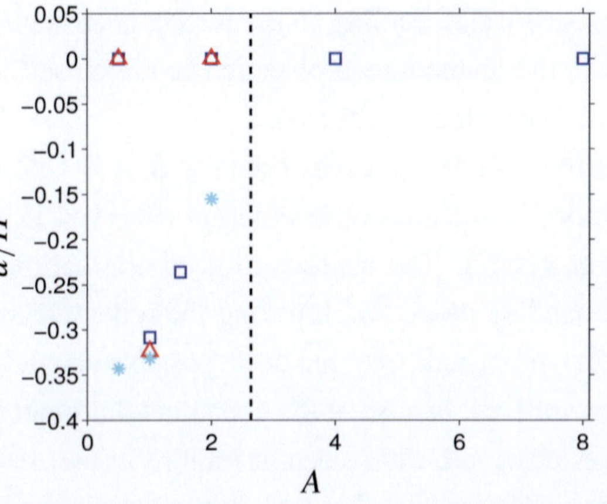

Figure 4.39: Submerged distance d from the free-slip plane $y/H = 1$ to the location of the maximum mean streamwise velocity $\langle u \rangle$ along the duct mid-span z/H. The distances are normalised by H and express as function of aspect ratio A. Coloured symbols indicate different bulk Reynolds number Re_b: □, $Re_b = 2205$; △, $Re_b = 3000$; $*$, $Re_b = 5000$. Black dashed line indicates the critical aspect ratio $A_c = 2.5$ suggested by Nezu and Rodi (1985).

Finally, the spanwise distribution of the dip-phenomenon submerged depth are summarised for each of three considered Reynolds number, and compared with the dip-phenomenon models of Wang et al. (2001) and Yang et al. (2004) in Figure 4.40. It is visible that for all Reynolds numbers, those models predict the submerged depth particularly well where the distance to the closest side-wall approximately equals to the duct full-height (i.e. $z''/H \approx 1$). Moreover, it was found in the highest Reynolds number flow that the model of Yang et al. (2004) continues to predict the spanwise dip-phenomenon profile well all the way up to $z''/H = 2$. This discussion on the dip-phenomenon model performance, however, should be considered as preliminary, since the Reynolds number dependence in the dip-phenomenon profile around $z''/H = 2$ could not be studied as mentioned before. Once again, further investigations based on the simulations with higher Reynolds number at this aspect ratio $A = 2$ (and beyond) should be performed in the future.

4.5 Conclusion

In total 23 open duct, and 3 additional closed duct simulations were performed for investigating the aspect ratio dependence in the mean secondary flow motions, with Reynolds number from the marginal to the fully-developed regime, over a range of aspect ratio (A) from 0.5 to 8.

It was found that the minimal sustainable Reynolds number limits are function of aspect ratio as in the closed duct counterpart, which were determined at $Re_b = 1900$ ($Re_\tau \approx 135$) for $A = 0.5$, and $Re_b = 1080$ ($Re_\tau \approx 74$) for $A = 2$. Unlike the minimal Reynolds number open duct flow with $A = 1$, the determined minimal flow exhibit: a unique four-vortex secondary flow pattern where only the duct side-walls host one low-velocity streak each for $A = 0.5$; and a two-vortex state where only one low-velocity streak is present on the bottom-wall boundary for $A = 2$. In both marginal flow configurations, the mechanism of the disappearance of the velocity streaks from the no-slip walls was shown as a consequence of the boundary edge lengths in wall units being too short to host a minimum set of the near-wall turbulent structures (i.e. one low-velocity streak franked by a pair of counter-rotating quasi-streamwise vortices). In such scenario, the turbulent structures cannot be sustained near those restrictive walls, leading to the turbulent structures to be detached from the walls of concern, similar to the phenomenon observed in the closed duct transitional turbulence with large aspect ratios (cf. Takeishi et al. (2015)).

In case of the marginal turbulence in the narrow duct with $A = 0.5$, it was shown that the flow does not exhibit any temporal-alternation of orientation observed in the marginal square closed duct flow of Uhlmann et al. (2007). The absence of such alternation was conjectured to be due to the imposed free-slip boundary condition, breaking the reflective symmetry in y-direction and favouring the observed four-vortex state over the other possible secondary flow arrangements with the same numbers of mean vortices. In case of the open duct marginal flow with $A = 2$, some kind of temporal-alternation was observed, whose details require further investigations.

In all secondary flow patterns observed so far, both in the marginal and the fully-developed open duct turbulence, the mean secondary flow vortices in the mixed-boundary corners have the same sense of rotation, which corresponds to the direction of the inner secondary flow. This phenomenological consistency was associated with the vortex sorting hypothesis discussed in the previous chapter. In other words, the unfavoured vortices, determined by the vortex sorting mechanism, cannot be stably existed in the mixed-boundary corners to appear in the mean flow statistics.

All those newly-discovered secondary flow patterns as well as the observed preference in the rotational directions of the mean flow vortices for different cross-sectional locations, are considered to be crucial information for our future study to search for open duct non-linear invariant solutions.

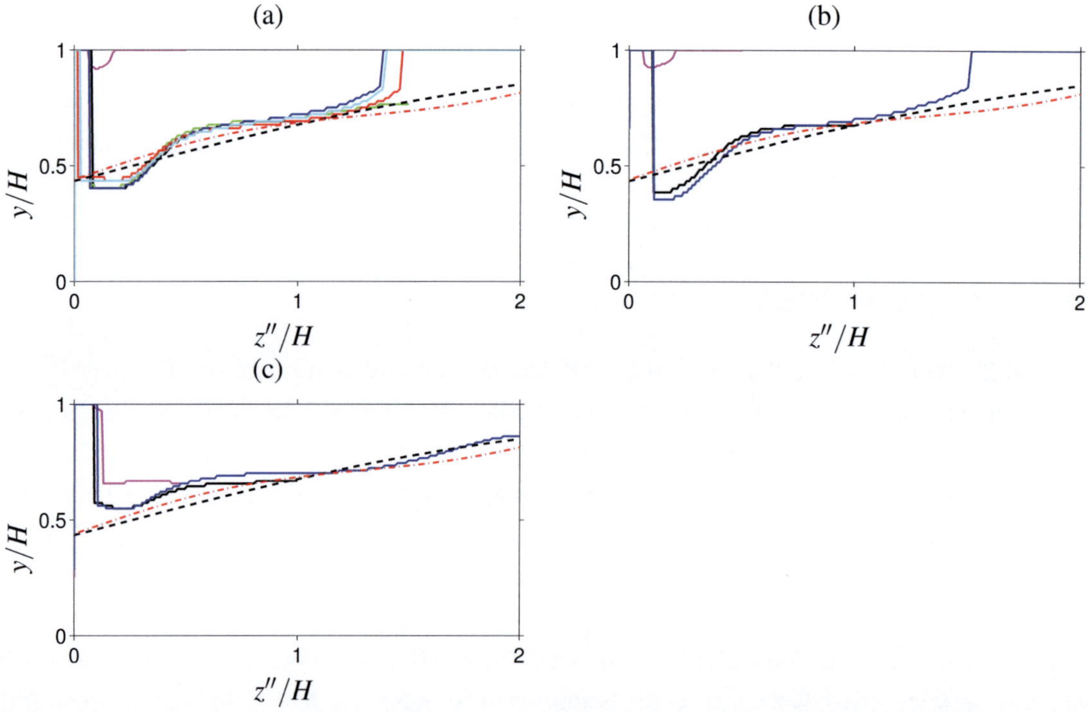

Figure 4.40: Locations of the maximum mean streamwise velocity as function of $z''/H = (AH - |z|)/H$ (black solid line). (a) $Re_b = 2205$, (b) $Re_b = 3000$, (c) $Re_b = 5000$. Colours of the solid lines represent different aspect ratio: magenta, $A = 0.5$; black, $A = 1$; blue, $A = 1$; green, $A = 1.5$; blue, $A = 2$; cyan, $A = 4$; red, $A = 8$. Red dotted dash and black dashed lines indicate the predicted values by the models of Wang et al. (2001) and Yang et al. (2004) respectively.

Subsequently, the fully-developed open and closed duct flow with $Re_b = 2205$, and a range of aspect ratio up to 8 were investigated.
The earlier experimental and numerical observations that the duct secondary flow is confined in the vicinity of the side-walls even in large aspect ratio ducts, were confirmed in a various aspects, including: the near-wall velocity streak locations; the secondary flow streamfunction and streamwise vorticity extrema locations; and the depth-averaged secondary flow intensity distribution in spanwise direction.
Through the examination of the depth-averaged secondary flow intensity distribution, it was also found that the averaged intensity becomes independent of aspect ratio, if we further integrate the averaged intensity in spanwise direction only over the distance $H(h)$ from the duct side-walls.
Moreover, it was found for both open and closed duct flow, that the secondary flow intensity decays exponentially as the distance from the nearest side-wall increases, approaching asymptotically to some constant decay rate as aspect ratio increases. For the range of aspect ratios considered here, the decay constant for the open duct flow with $A = 8$ was found at $\lambda \approx 0.5$, whereas the constant for the closed duct flow with $A = 8$ was found at $\lambda \approx 0.7$. The possible Reynolds number dependence in the decay properties is remained as a topic of future research.
Finally, a consequence of the velocity dip-phenomenon was investigated. To our knowledge, it was observed for the first time that there exists Reynolds number dependence in the critical aspect ratio for the global dip-phenomenon (i.e. the maximum streamwise velocity location is observed below the free-slip plane even at the duct mid-span). Our results demonstrated that in case of low Reynolds number flow, the dip-phenomenon disappears even if the duct aspect ratio is smaller than the widely-accepted critical value of $A_{\text{crit}} \approx 2.5$. At the time of writing, the observed Reynolds number dependence in the dip-phenomenon is conjectured as a consequence of the reach of the oblique band-like low-velocity structures observed exclusively in the region directly below the free-slip plane. Further structural and dynamical investigations on this topic is required to conclude the validity of the conjecture.

Chapter 5

Time-resolved evolution of turbulent coherent structures in open duct flow

5.1 Introduction

In the preceding chapters, the unique secondary flow patterns appearing in the open duct flow have been analysed chiefly based on their mean flow statistics averaged in time and in space along the homogeneous streamwise direction. Some of the geometrical aspects of the mean secondary flow (e.g. Reynolds-number scaling of the locations of the streamwise vorticity peaks) were explained by the numbers and arrangements of the near-wall turbulent coherent structures, such as high-/low-velocity streaks and quasi-streamwise vortices.

Consequently, a strong link between the mean flow statistics and the preferential locations of the instantaneous turbulent coherent structures was identified by means of various coherent structure eduction techniques applied upon a set of instantaneous flow fields. All of those coherent structure techniques extract some kind of structural information from the instantaneous flow fields independently from each other, and the statistics of interest are accumulated via standard ensemble averaging operations.

Whilst those *stationary* coherent structure analysis provided us some meaningful insights into the picture of those preferentially-located coherent structures being responsible for the mean secondary flow, there exists one major limitation in those techniques considered so far that the *dynamical* aspects of the educed coherent structures have been missed until this point. However, such dynamical information of the near-wall coherent structures are highly valuable, for example, to validate our hypothesis that the interactions between the quasi-streamwise vortices and the free-slip boundary form a vortex sorting mechanism that is only active within the immediate vicinity of the free-slip plane. Since the sorting dynamics is expected to be a key element that is responsible for the appearance of the unique open duct secondary flow patterns, i.e. inner/ outer secondary flow manifested within the mixed-boundary corners and along the free-slip plane, quantifications of such dynamical phenomenon should be addressed within this study.

Furthermore, it is equally (if not more) important to underline the similarities and discrepancies between the open duct vortex dynamics and the equivalent dynamics in the canonical plane channel

and boundary-layer flow that have been studied extensively over the last years. We expect some significant differences in the averaged pictures of the vortex dynamics being not only influenced by the distance from the top or bottom-wall boundaries, but also by the distance from the side-walls and the associated corners, as the other statistical quantities discussed so far. Note that similar studies of the open channel vortex dynamics have not been performed previously.
The main objective of this chapter is, therefore, to propose a new approach to educe and track the three-dimensional quasi-streamwise vortices embedded within a highly-resolved temporal series of the open duct direct numerical simulation data. Note despite the generality of the eduction and tracking techniques that in theory allows us to consider any kind of turbulent coherent structures (e.g. high-/low-velocity streaks, vortices defined by Q-criterion (Hunt et al. (1988)), or by λ_2-criterion (Jeong and Hussain (1995))), we shall restrict ourselves to the structures defined by closed iso-value surfaces of the streamwise vorticity ω_x in this study. Such selection can be explained by our primary interest in the mean secondary motions in open duct flow, and can be supported by our earlier findings that the secondary motions are the statistical footprints of the quasi-streamwise vortices. Such vortices can be preferentially extracted by utilising the iso-value surface of the streamwise vorticity. The configuration of the numerical data set was chosen to be our reference open duct case with the bulk Reynolds number at 2205 whose aspect ratio is set at unity. The resulting individual trajectories shall be then summarised based on the conditions of our interests, such as comparisons between the vortices which exist in the vicinity of the free-slip plane and the bottom-wall boundary, as well as their temporal evolution (or fate).
The rest of this chapter is organised as follows. First, the key aspects of the employed numerical data set will be discussed together with the representativeness of the data set to the fully-converged mean secondary flow. Then, the considered vortex eduction technique will be explained and again the significance of the educed vortices to the overall mean flow fields will be examined. Finally, the new vortex tracking technique will be introduced, and the outcomes of the tracking analysis will be discussed at the end, being followed by the overall conclusion.

5.2 Numerical dataset

In total 3000 instantaneous flow fields of the open duct flow with $Re_b = 2205$ ($Re_\tau \approx 150$) and $A = 1$ were generated by means of the pseudo-spectral direction numerical simulation code utilised throughout this dissertation work. The case set-up of the simulation is identical to the one presented in Chapter 3, with the spatial resolutions of $\Delta x^+ \approx 14.8$ and $\max\{\Delta y^+\} \approx \max\{\Delta z^+\} \approx 2.46$, whereas the CFL limit was kept below 0.3. The instantaneous flow fields are separated by $\Delta t_s^+ \approx 1.5$, where Δt_s^+ is the temporal separation between each field normalised by the viscous time unit $\delta t_\nu = \nu / u_\tau^2$. The resulting temporal series spans over $\approx 4500\delta t_\nu$, which corresponds to approximately 427 bulk time units. The key numerical parameters of the simulation are summarised in Table 5.1.
The sufficiency of the instantaneous data sample was evaluated by comparing the ensemble-averaged streamwise vorticity profile from the dataset $\langle\omega_x\rangle_{\text{inst}}$, and the fully-converged mean streamwise vorticity profile accumulated over ≈ 13000 bulk time units $\langle\omega_x\rangle$ (cf. Table 3.1). As

A	Re_b	Re_τ	$M_x \times M_y \times M_z$	Δx^+	$\max\{\Delta y^+, \Delta z^+\}$	Δt_s^+	T_s^+	$T_s u_b/H$
1	2205	150	$256 \times 97 \times 193$	14.8	2.46	1.5	4500	427

Table 5.1: Parameters of the current open duct numerical data set employed for vortex dynamics study.

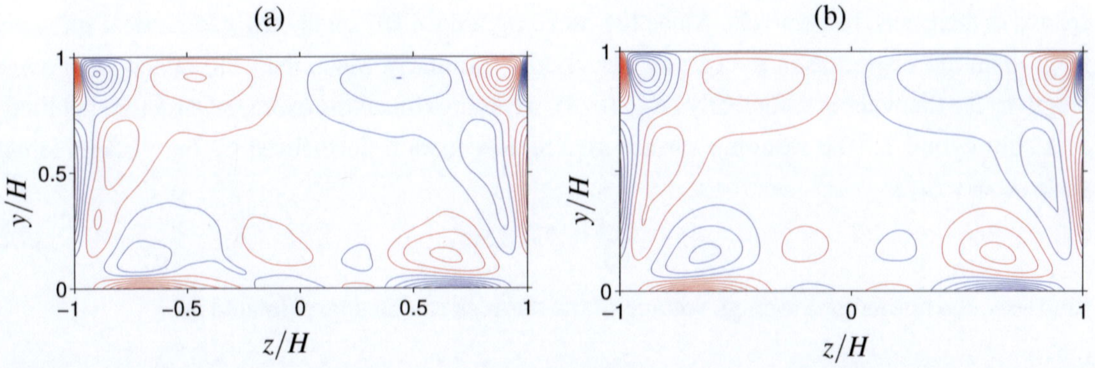

Figure 5.1: (a) Ensemble-averaged streamwise vorticity contour $\langle\omega_x\rangle_{\text{inst}}$ from 3000 instantaneous fields accumulated over an interval of 427 bulk time units. (b) Mean streamwise vorticity contour $\langle\omega_x\rangle$ averaged over 13000 bulk time units. For both cases, blue contour lines indicate -0.1(-0.1)-0.9 times the maximum value of the mean streamwise vorticity $\langle\omega_x\rangle$ at 0.3016. Red contour lines indicate 0.1(0.1)0.9 times the same maximum value. Bulk Reynolds numbers are at 2205 and aspect ratios are set at unity.

it is evident in Figure 5.1, the both profiles show a great qualitative agreement that confirms the sufficiency of the data set for the purpose of this analysis.

5.3 Vortex eduction

In this section the new vortex eduction technique employed in the current study will be explained. The eduction technique of concern here can be summarised in the following two-step algorithm form: (i) eduction of iso-value structures for a given threshold value, based on the surface triangulation method; (ii) elimination of insignificant microscopic structures according to their surface area or enclosed volume for a given threshold value.

Before proceeding any further, a significant contribution from Dr. Kidanemariam by providing a key MATLAB routine employed in the step (i) should be acknowledged at this point. The provided routine accepts a field value reside on a three-dimensional grid points, say ω_x in this study, then extracts the surface geometry of individual structures as closed iso-value surface defined by a set of vertices and faces for a given threshold value. Based on those vertices and faces, the three-dimensional coordinate of the centroid, surface area and enclosed volume of each structures are computed.

The educed vortices are then filtered by a user-specified surface-area or volumetric threshold, depending on our needs (step (ii)). Such filtering operation is necessary to retain majority of statistically-significant structures whilst maintaining the level of the computational effort of the forthcoming tracking procedure within a feasible limit. As a guideline for the surface-area and volumetric threshold values, the average size of the near-wall vortices educed by λ^2-criterion and studied by Jeong et al. (1997) were consulted. Note that the data employed in their coherent structure analysis are from the low Reynolds number plane channel direct numerical simulation by Kim et al. (1987), whose $Re_\tau \approx 180$ is equivalent to the current open duct dataset (i.e. $Re_\tau \approx 150$). Jeong et al. (1997) reported that the average size of the educed near-wall vortices is 25 wall units in diameter, 200 wall units in streamwise length, inclined by 9° and 4° in the wall-normal and spanwise directions respectively. Since the inclining angles are small, any geometrical influence from them are neglected in the current discussion. Similarly, since the ratio of the streamwise length to the diameter is sufficiently large (= 8), we approximate the averaged structure geometry as a slim cylinder. The resulting surface area approximation normalised by the viscous length scale $\delta_\nu = \nu/u_\tau$ is

$$\frac{S_{\text{est}}}{\delta_\nu^2} \approx 5312.5\pi \,. \tag{5.1}$$

Similarly, the normalised average volume of the vortices can be approximated as:

$$\frac{V_{\text{est}}}{\delta_\nu^3} \approx 31250\pi \,. \tag{5.2}$$

By considering the fact that the educed ω_x structures are only the streamwise extension of the near-wall vortices (or "legs" of hairpin vortices), a relatively small fraction (say $\leq 10\%$) of the approximated average values should be employed as the threshold values. In the current investigation, only the filtering by the surface area of the educed structures are considered for simplicity. For the purpose of evaluating the influence of our choice of the threshold parameters, a set of three vorticity threshold values were chosen based on the maximum intensity of $\langle\omega_x\rangle_{\text{rms}}$ (cf. Figure 3.13(d)), namely $\omega^+_{x,\text{th}} = \pm[0.08(27\%),\ 0.16(53\%),\ 0.24(80\%)]$. Note the specified percentages are in comparison with the maximum intensity of $\langle\omega_x\rangle_{\text{rms}}$. Similarly, a set of four surface-area threshold values were chosen, which are $S_{\text{th}}/S_{\text{est}} = [0, 0.051, 0.067, 0.08]$. Note that $S_{\text{th}}/S_{\text{est}} = 0$ means no filtering was applied upon the educed structures.

First, let us consider the influence of the threshold parameters in the number of detected structures. Figure 5.2(a) shows the threshold dependency of the average number of the structures which are detected in a snapshot $\langle N_\text{v}\rangle$. For all the surface-area threshold values, the number first increases and then decreases as the vorticity threshold value $\omega_{x,\text{th}}$ increases. This particular trend indicates that the lowest threshold value ($\omega^+_{x,\text{th}} = \pm 0.08$) is too low so that some of the individual structures are connected artificially, whereas the highest value ($\omega^+_{x,\text{th}} = \pm 0.24$) is too restrictive to detect some of the structures with weak intensities. A radical reduction in the number of the structures can be observed by applying the least restrictive finite surface-area filter of $S_{\text{th}}/S_{\text{est}} = 0.051$ to the original set of the vortex structures, whereas the following changes caused by the additional filtering operations seem more moderate.

Next, we shall investigate the parameter dependency in the volumes of the educed vortex structures. Figure 5.2(b-d) show the threshold dependency of: (b) the total, (c) maximum and (d) average volume of the structures which are detected in a snapshot, based on the volume of the individual structures V_i. Note the index $i = 1 \ldots N_\text{v}$ represents the individual structures reside in a snapshot. Whilst the general trend of monotonic decrease of all three volumetric values with increased vorticity threshold satisfies our expectation, it is important to notice that the total volume (Figure 5.2(b)) are affected only slightly, even after the most restrictive filtering. This observation

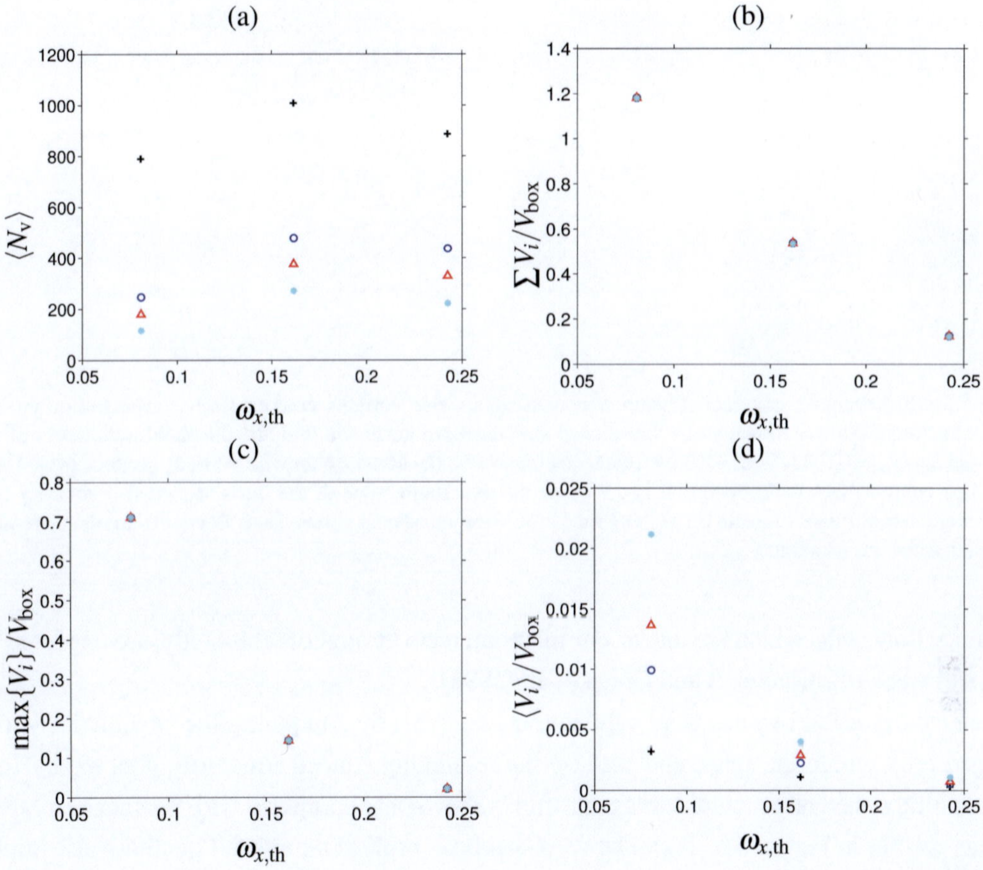

Figure 5.2: (a) Number and (b-d) various volumes of the detected structures as function of streamwise vorticity threshold value $\omega_{x,\rm th}$. Definitions of (b-d) are: (b) total, (c) maximum and (d) averaged volume normalised by the numerical box volume $V_{\rm box} = L_x L_y L_z$. Different coloured symbols indicate four different surface-area threshold values normalised by estimated average surface area: $+$, $S_{\rm th}/S_{\rm est} = 0$ (unfiltered); $\circ$, $S_{\rm th}/S_{\rm est} = 0.051$; $\triangle$, $S_{\rm th}/S_{\rm est} = 0.067$; $*$, $S_{\rm th}/S_{\rm est} = 0.084$. Averaged over 10 instantaneous fields of open duct with $Re_b = 2205$ and $A = 1$, which are separated by ≈ 150 viscous time units.

implies that large number of filtered (order of several hundreds) structures are indeed insignificant in their sizes and have virtually no contributions to the total volume of the detected structures. Note also that the total volume of the structures educed with $\omega^+_{x,\rm th} = \pm 0.08$ exceeding the volume of the hosted numerical box $V_{\rm box}$ should be again a consequence of the vorticity threshold value being too low. At such low vorticity threshold value, some of the structures are artificially connected and often form largely unclosed surfaces that the volume computation cannot be performed correctly.

Naturally, the maximum volumes are not affected by any of the surface-area filtering operations that only target the smaller end of structures, whereas the averaged vortex volumes are significantly increased by the filtering operations by eliminating the microscopic structures.

Based on the above parameter-dependency studies, we selected $\omega^+_{x,\rm th} = \pm 0.16$ and $S_{\rm th}/S_{\rm est} = 0.067$ as our choice of the threshold parameters. Whilst the selection of the vorticity threshold was straightforward by choosing the value with the highest peak in Figure 5.2(a), the intermediate value of the surface-area threshold was chosen with some safety in mind. The choice was made by taking into account our earlier observations that those vortices responsible for the mean inner

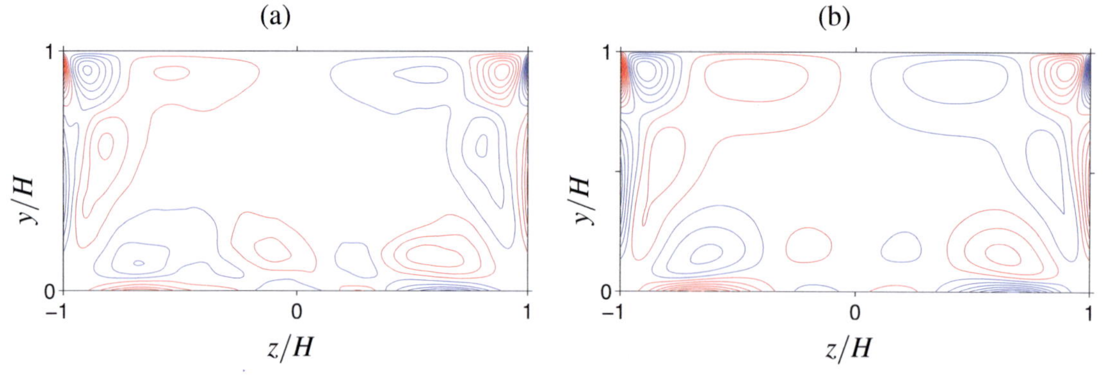

Figure 5.3: (a) Ensemble, coherent structure-averaged streamwise vorticity contour $\langle\omega_x\rangle_{cs}$, averaged only over the educed structures' cross-sectional area. The educed structures are generated with the threshold parameters of $\omega_x^+ = \pm 0.16$ and $S_{th}/S_{est} = 0.067$, from 3000 instantaneous snapshots. (b) Mean streamwise vorticity contour $\langle\omega_x\rangle$. For both cases, blue contour lines indicate -0.1(-0.1)-0.9 times the maximum value of the mean streamwise vorticity $\langle\omega_x\rangle$ at 0.3016. Red contour lines indicate 0.1(0.1)0.9 times the same maximum value. Bulk Reynolds numbers are at 2205 and aspect ratios are set at unity.

secondary flow cells, which are one of our main interests throughout this study, are expected to be small and weak (cf. Chapter 3 and Grega et al. (1995)).

The selected threshold values were validated *a posteriori* by computing the streamwise vorticity averaged only within the space enclosed by the remaining educed structures, denoted by $\langle\omega_x\rangle_{cs}$. The resulting coherent structure average profile is shown and compared with the mean streamwise vorticity profile in Figure 5.3. It can be seen that both profiles agree well qualitatively, implying that the educed and filtered structures contain the quantities of our interest.

Preferential locations of the educed vortex centroids were also computed and presented in Figure 5.4. Note that the contributions from the near-wall structures whose centroids are located less than 4 viscous length scale away from the nearest no-slip walls had to be removed to highlight the contributions from the other structures residing elsewhere. Those eliminated structures that reside within the viscous sub-layer are the mirrored signatures of the neighbouring vortices exist directly above. Their statistical contributions appear as sheet-like wall-attached peaks in the mean vorticity profile, which exist between the secondary vorticity peaks and the closest no-slip walls. The current observation that the contributions from the sub-layer structures are predominant with respect to the equivalent ones in the other parts of the duct cross-section indicates a geometrical nature of those near-wall structures: being short in streamwise direction whilst existing in relatively large numbers in certain preferential locations.

In contrast, the probability peaks corresponding to the inner secondary flow cells are not as predominant as it seemed in the coherent structure analysis based purely on cross-sectional slice information (cf. Figure 3.13(c), for instance). This observation implies that those vortices located within the mixed-boundary corners are highly elongated in streamwise direction, which result one structure being counted many times in the two-dimensional probability analysis, whilst the true number of three-dimensional coherent structures is reflected in the current analysis.

Additional structural information can be deduced from the cross-sectional distribution of the averaged volume of the educed structures $\langle V_i\rangle$. The spatial distributions of those volumetric information are characterised by their centroid locations $(y_c, z_c)_i$ and normalised by the numerical box

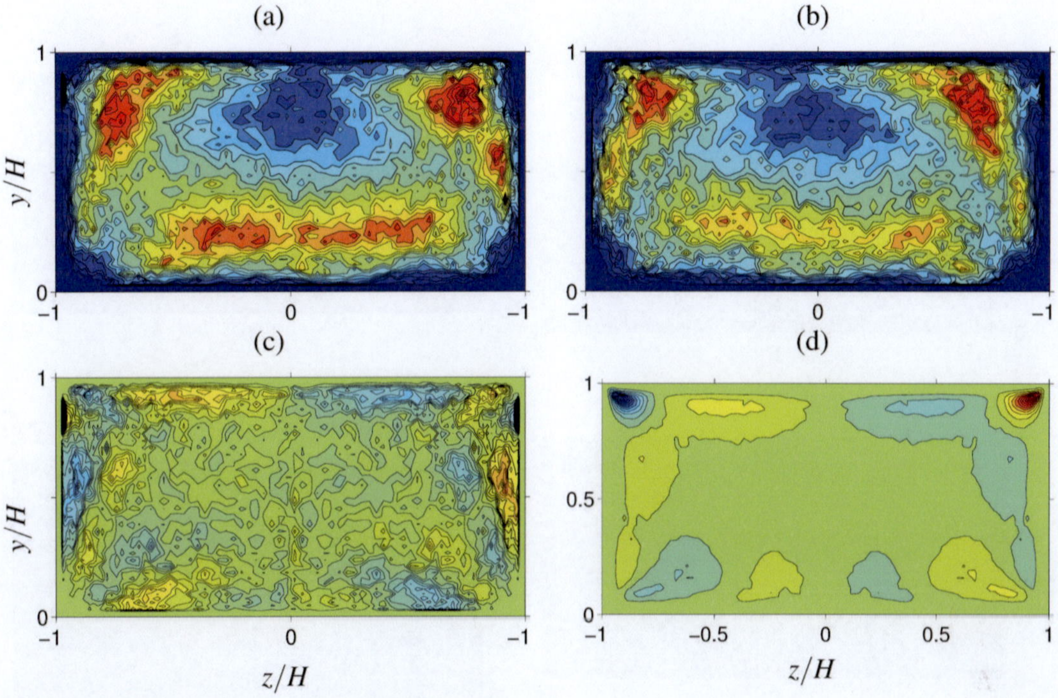

Figure 5.4: Probability of occurrence of educed vortex centroids for the open duct case with $Re_b = 2205$ and $A = 1$. (a) vortices with *positive* streamwise vorticity; (b) *negative* streamwise vorticity; (c) difference of (a) and (b), being symmetrised about the duct mid-span. Contributions from the near-wall shear structures were eliminated by neglecting the detected structures with their centroid locations closer than 4 viscous units from the nearest no-slip wall. The educed structures are generated with the threshold parameters of $\omega_x^+ = \pm 0.16$ and $S_{th}/S_{est} = 0.067$, from 3000 instantaneous snapshots. (d) Probability of occurrence of vortex centres for the open duct case with $Re_b = 2205$, detected by the technique proposed by Kida and Miura (1998) (same as Figure 3.13(c)). Reproduced for comparison.

volume V_{box} (cf. Figure 5.5). The two-dimensional cell-based volume averaging operation was performed by projecting the centroid locations onto a 10-by-20 equidistant grid.

It can be seen that the largest vortices are located above the bottom-wall boundary around the duct mid-span, whilst the side-wall maxima exists somewhere above the duct mid-height. Note that those peak locations correspond fairly well with the stagnation points of the mean secondary flow patterns along the no-slip boundary edges, except for the corner regions. Such observation is consistent with our earlier finding that the mean open duct secondary flow patterns are, apart from the inner secondary flow, outer-scaling (large-scale) phenomenon. In other words, we expected some signatures of the large-scale structures associated with the large-scale secondary motions. On the other hand, both the solid-solid and mixed-boundary corners host small-scale structures where such large-scale motions are absent. The small-scale nature of the mixed-boundary corner vortices confirms the expectations based on our earlier investigations.

Comparison against the estimated average volume defined in Equation 5.2 is also interesting (cf. Figure 5.5(d)). It turns out that the value on the bottom-wall bisector is approximately ten times larger than the estimated volume, alongside the side-wall peak values being approximately 4 times larger. The results in the other regions, by contrast, seem to be fairly consistent with the estimated volume V_{est}. An example of such large-scale vortex structure located near the bottom-wall bisector is presented in Figure 5.6.

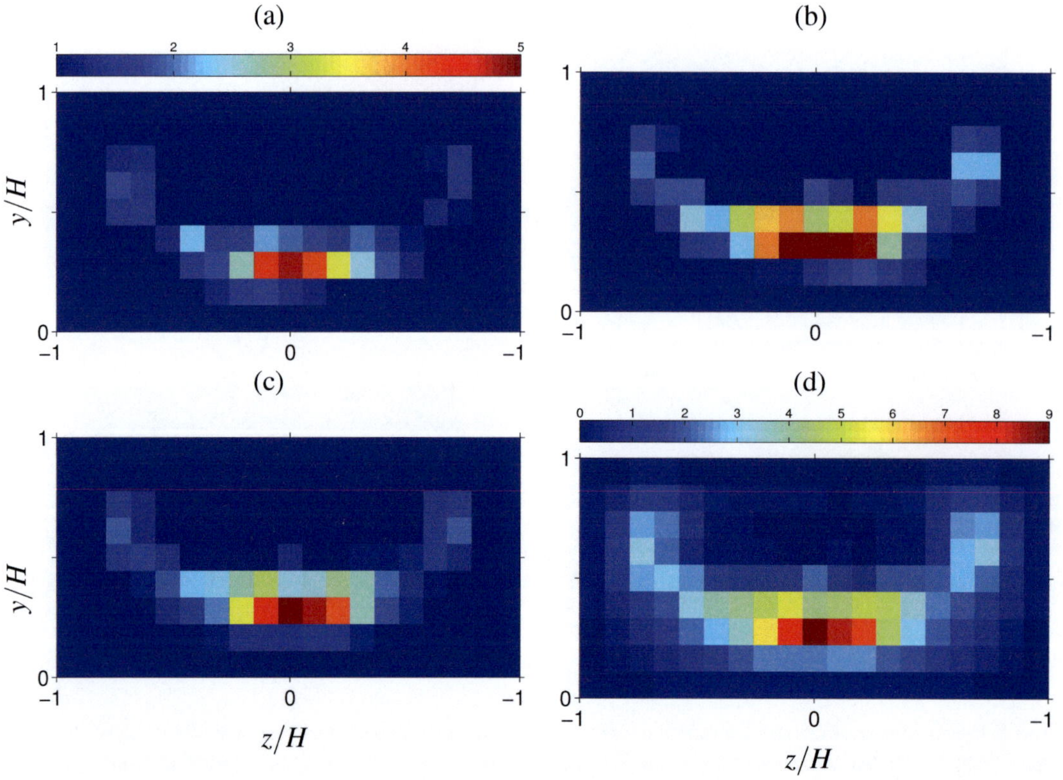

Figure 5.5: Spatial distribution of averaged educed vortex volume, normalised by (a-c) numerical box volume V_{box} (the scale of the numbers on the colour bar are $\times 10^{-3}$), or (d) estimated vortex volume V_{est}. The volumes of the individual structures are associated with the centroid locations and were averaged over square boxes enclosed by 10-by-20 equidistant grid points (not CGL points). The educed structures are generated with the threshold parameters of $\omega_x^+ = \pm 0.16$ and $S_{\mathrm{th}}/S_{\mathrm{est}} = 0.067$, from 3000 instantaneous snapshots of the open duct with $Re_b = 2205$ and $A = 1$. (a) vortices with *positive* streamwise vorticity; (b) *negative* streamwise vorticity; (c) average of (a) and (b). (d) same as (c) but normalised by V_{est}.

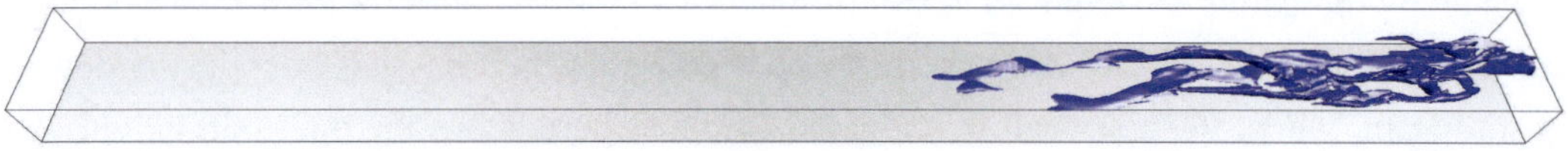

Figure 5.6: Example of large-scale vortex structure in open duct flow with $A = 1$ and $Re_b = 2205$; flow direction is from left to right. The vortex was educed by $\omega_x^+ = -0.16$.

Those volumetric measures are by nature largely depend on the employed vortex criterion as well as the vorticity threshold value. In any case, it is important to report the inhomogeneous distribution of the size of the vortices in the open duct flow, which has not been studied thoroughly in the past to the best of our knowledge. To draw more rigorous conclusions in this volumetric aspect of the open duct flow, especially in the context of the large-scale motions that have been studied extensively for other turbulent canonical flow with much higher Reynolds numbers, it is necessary to make a use of the three-dimensional eduction techniques that designed for such multi-scale phenomenon. One promising candidate for this kind of analysis is the definition of the vortex clusters studied by del Álamo et al. (2006), which is based on the discriminant of the velocity gradient tensor. Further investigations based on such techniques applied upon higher Reynolds number open/closed duct flow are planned for the future studies.

5.4 Temporal evolution of educed vortices

5.4.1 Vortex tracking method

Here we shall explain the proposed method of temporal tracking of the educed vortices, which is primarily based on the centroid location of each structure, but additionally on their overlapped volumes with the structures of the subsequent snapshot. Note that similar time-tracking criteria have been employed independently, for instance, by Lehew et al. (2013) and Lozano-Durán and Jiménez (2014) respectively.

The proposed hybrid criterion was developed in order to achieve a compromise between the computationally efficient nature of the centroid-based method, and the robustness of the overlapped-volume method that is computationally more expensive. However, the attractive capability of the volumetric technique of identifying occasional splitting and merging events of the vortices had to be compromised. For the sake of clarity, detailed explanations on the limitation of the current time-tracking method will be offered with appropriate examples in the following part of this section.

Before proceeding any further, let us define some key terminologies necessary to explain the tracking method. We shall indicate any field values $\phi(x,y,z,t)$ in an arbitrary snapshot with the superscript n, i.e. $\phi^n = \phi(x,y,z,t^n)$, where t^n is a discrete point in time corresponding to the snapshot of concern. Consequently, the field value in the subsequent snapshot at $t^{n+1} = t^n + \Delta t_s$ is denoted as ϕ^{n+1}.

Prior to the time-tracking, the centroid locations of every educed structure i are computed as mentioned earlier. Additionally, the mean velocity components on the computed centroid locations ($\langle u(y_c,z_c)_i\rangle$, $\langle v(y_c,z_c)_i\rangle$ and $\langle w(y_c,z_c)_i\rangle$) need to be computed by interpolation from the simulation grid. Furthermore, the signed threshold values that have been used to educe those streamwise vortices are also stored to identify their sense of rotation. In the following, therefore, only the vortices with the same sense of rotation will be connected.

Upon the initiation of the time-tracking procedure, every structure at t^{n+1}, which is denoted by the index $j = 1\ldots N_v^{n+1}$, is translated backwards in space using the local mean velocity, viz.

$$[\tilde{x}_c,\tilde{y}_c,\tilde{z}_c]_j = [x_c,y_c,z_c]_j - \Delta t_s[\langle u(y_c,z_c)_j\rangle,\langle v(y_c,z_c)_j\rangle,\langle w(y_c,z_c)_j\rangle] \tag{5.3}$$

where $[\tilde{x}_c,\tilde{y}_c,\tilde{z}_c]_j$ denotes the shifted centroid coordinates of every structure j. This shifting procedure is mostly consistent with the one applied in Lozano-Durán and Jiménez (2014), and based on the earlier findings in the turbulent plane channel flow that the advection velocities of the small-scale turbulent fluctuations are approximately equal to the local mean velocities (cf. Kim and Hussain (1993), del Álamo and Jiménez (2009)). Note the method proposed by Lozano-Durán and Jiménez (2014) translates every vertex that defines the skeletons of their turbulent structures independently, based on the local mean velocity fields interpolated onto the vertices. In this way, the morphing of the structures resulted from the local velocity gradient in the wall-normal direction can be taken into account, in contrast to the current method that simply translates the whole structures around their centroids without any deformations. Although this part of their procedure

seems promising, it had to be omitted for the current implementation again due to the associated high computational effort and storage requirements.
Some exceptions should be mentioned for the advection velocities of the near-wall small-scale structures below $y^+ \approx 15$, as well as the large-scale structures that both exist in the plane channel turbulence. Whilst the small-scale near-wall structures travel with an approximately constant velocity at $\approx 11u_\tau$, those large-scale structures were found to travel with a constant velocity that is approximately equivalent to the bulk flow velocity throughout the entire channel domain (cf. del Álamo and Jiménez (2009)). Despite the physical importance of those quantities in the current flow configuration, we need to proceed without any further discussion on this topic at this point, since it is simply out of the scopes of the current study and technique. However, further investigations on this topic should be carried out separately in the near future, preferably by developing a new spectral-based method similar to the one employed by del Álamo and Jiménez (2009).
Let us now consider a particular vortex at t^n, denoted by the index $\tilde{i}$. The ultimate purpose of this part of the tracking procedure is to find its successor in the subsequent snapshot at t^{n+1} (the succeeding vortex is denoted by $\tilde{j}$).
After shifting every vortex structure at t^{n+1} backwards in space based on the mean velocity field at the centroids, we construct a set of the position difference vectors $\boldsymbol{r}_{\tilde{i},j}$, which are defined with respect to the centroid location of the vortex $\tilde{i}$ at t^n and its distance to the centroid locations of every vortex j. As a result of the backward shifting, the position difference vectors should now be solely the product of the geometrical morphing of those vortices over the time interval Δt_s. Subsequently, the position difference vectors are sorted according to their magnitude in ascending order, and the vortex associated with the first element of the sorted vector is selected as a succeeding candidate $\tilde{j}$. Furthermore, the magnitude of $\boldsymbol{r}_{\tilde{i},\tilde{j}}$ is compared against the predefined searching radius r_th, and if $|\boldsymbol{r}_{\tilde{i},\tilde{j}}| < r_\mathrm{th}$, then a connection between $\tilde{i}$ and $\tilde{j}$ is established (cf. Figure 5.7(a,b) for illustrative examples). Note that the value of r_th is designed to be function of V_est, based on a consideration that the larger structures are subjected to the stronger velocity shear and should therefore result a larger displacement of the centroid location by higher degree of morphing. The definition of r_th is as follows:

$$r_\mathrm{th} = \alpha_r (V_\mathrm{est})^{1/3} \tag{5.4}$$

where α_r is a user-specified constant parameter that was set at 0.5 throughout the current study.
In case of more radical changes of the centroid locations that results $|\boldsymbol{r}_{\tilde{i},\tilde{j}}| > r_\mathrm{th}$, which is often associated with vortex splitting/merging events, the overlap volume between the vortex $\tilde{i}$ and $\tilde{j}$ needs to be assessed. In order to compute the overlapped volume, their vertices are interpolated onto a dedicated separate three-dimensional grid with no variation in the grid spacing in each direction (in contrast to the simulation grid which has variations in two cross-stream directions). In this way, every cubical cell of the dedicated grid has an identical volume V_c that one simply needs to count those cells whose centroid locations are shared by the two structures in order to compute the overlapped volume. Furthermore, the spatial resolution of the dedicated grid, which is solely depend on the number of the grid points in each direction (i.e. N_x, N_y and N_z), can be adjusted according to the computational capacity of the users. For the purpose of the current

study, we selected the numbers of the grid points in each direction as a half of the numbers of the Fourier/Chebyshev-Gauss-Lobatto grid points utilised in the simulation.
Consequently, the overlapped volume $V_{\tilde{i}\wedge\tilde{j}}$ is defined as follows:

$$V_{\tilde{i}\wedge\tilde{j}} = \sum V_\mathrm{c},\ V_\mathrm{c} \in \{V_\mathrm{c} | V_\mathrm{c} \in \tilde{i} \wedge V_\mathrm{c} \in \tilde{j}\} \tag{5.5}$$

where V_c is the volume of one uniform grid cell. Finally, the ratio of $V_{\tilde{i}\wedge\tilde{j}}$ to the volume of the structure $\tilde{j}$ is compared against another user-specified threshold ratio α_v, that was set at 0.4 in this case. A connection between $\tilde{i}$ and $\tilde{j}$ is established if the condition $V_{\tilde{i}\wedge\tilde{j}}/V_{\tilde{j}} > \alpha_\mathrm{v}$ is met. Note that in case of splitting, only the large-enough child-structure that is closest to the parent structure in their centroid locations are considered as the successor, whilst the remaining child structures are recognised as independent newly-born structures to be tracked from that point in time (cf. Figure 5.7(b,c)). This limited capability of identifying the splitting events is one of the compromises made for this current method that was mentioned earlier.
If either the centroid distance or the overlapped volume criterion is met, then the time is advanced ($t^{n+1} \to t^n$) with $\tilde{j}$ becoming new $\tilde{i}$ (i.e. the two candidates are identified as one), and the tracking process continues. If none of the two conditions are met, on the other hand, the time-tracking of $\tilde{i}$ is terminated at that point in time and $\tilde{j}$ becomes a new track. Please refer to Figure 5.7(d) for an illustration of a case of loosing a track artificially.
Streamwise domain periodicity is taken into account continuously by checking the gap between $(x_\mathrm{c})_{\tilde{i}}$ and the streamwise domain limit L_x. In case of the gap is smaller than $0.5\Delta t_\mathrm{s}\langle u(y_\mathrm{c}, z_\mathrm{c})_{\tilde{i}}\rangle$, the centroid streamwise location is simply translated by L_x backwards before the successor searching routine is initiated.
To this end, it is important to note that the temporal tracking is performed in an iterative manner, by starting from all the vortices that are not marked as initial/succeeding structures previously. Similarly, those structures that have been tracked before are not considered to be candidates of the successors. In other words, all the educed vortex structures are marked as initial points or successors strictly only once, which implies to the incapability of detecting the merging events within the current framework. The limitation is considered to be acceptable for the current purpose of investigating the dynamical aspects of the vortices in the mixed-boundary corners and free-slip plane, since the merging events are usually associated with the inverse cascade to the turbulent kinetic energy that is not our interest here. In any case, incorporating the sophisticated graph-based structural identity management performed in Lozano-Durán and Jiménez (2014) should be considered in the future, in order to achieve more complete picture of the splitting/merging events in the open duct flow.

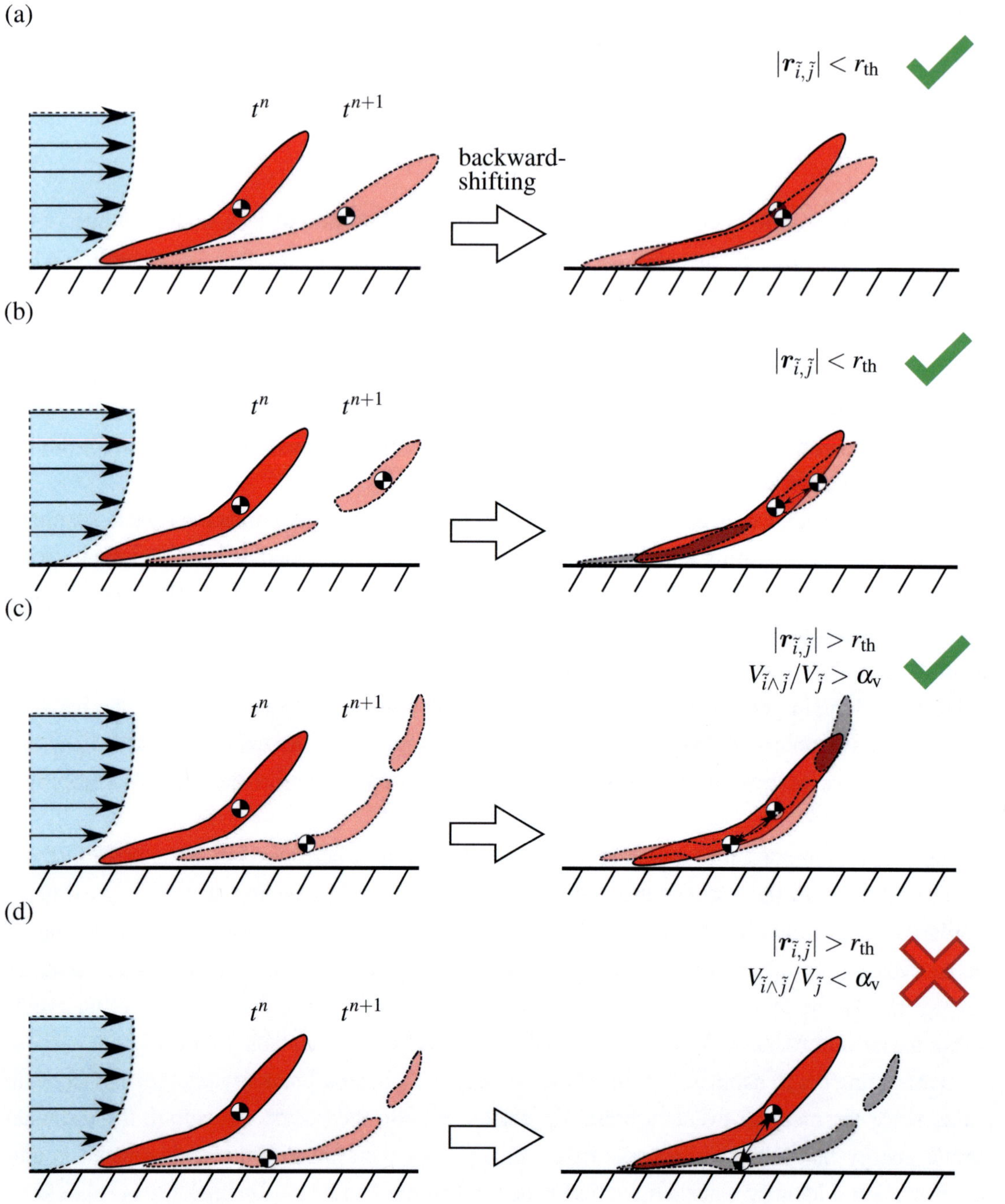

Figure 5.7: Schematics of four representative scenarios in the vortex tracking procedure, where: (a) case of successful vortex connection by the condition $|\boldsymbol{r}_{\tilde{i},\tilde{j}}| < r_{\text{th}}$ without splitting; (b) same as (a) with splitting; (c) case of successful vortex connection by the condition $V_{\tilde{i}\wedge\tilde{j}}/V_{\tilde{j}} > \alpha_{\text{v}}$; (d) case of unsuccessful vortex connection. Diagrams on the left and right sides are with and without backward shifting of the structures at t^{n+1} respectively. Red structures, the vortex $\tilde{i}$ at t^n; half-transparent red structures, the candidates (left) or the connected successor (right) $\tilde{j}$ at t^{n+1}; grey structures, independent structures without being connected. Only the closest candidates of $\tilde{j}$ have their centre of gravity indicated. The length of the dual-head arrows indicates $|\boldsymbol{r}_{\tilde{i},\tilde{j}}|$.

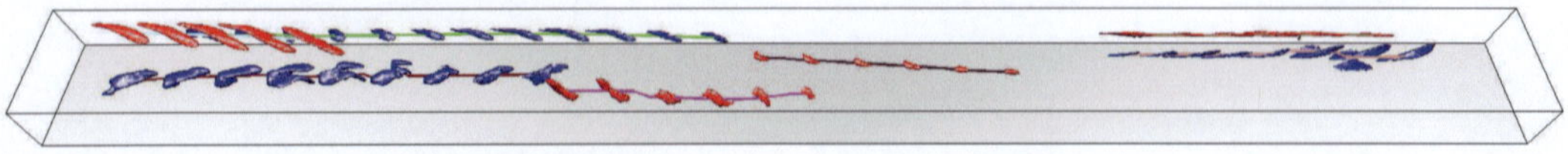

Figure 5.8: Vortex temporal-tracking result in open duct flow with $A = 1$ and $Re_b = 2205$; flow direction is from left to right. The image shows several arbitrarily selected structures over a time sequence covering approximately 8 bulk time units ($\approx 82.5\Delta t_v$), connected by solid lines. The colour of the structures indicates *positive* (red) and *negative* (blue) values of the streamwise vorticity thresholds employed to educe those structures.

5.4.2 Tracked vortex trajectories

A temporal-tracking of vortices was performed on the dataset described in Section 5.2, which consists of 3000 snapshots of the open duct with $A = 1$ and $Re_b = 2205$ ($Re_\tau \approx 150$), separated by $\Delta t_s^+ \approx 1.5$.

In total 124413 vortex trajectories were followed, including the 27342 trajectories (or rather points) that consist of vortex structures which have been identified only at singular point in time. Some of the tracked vortex trajectories are selected and presented in Figure 5.8 for the illustrative purpose.

We estimate the efficiency of the tracking method, which is at $\approx 78\%$, by assuming those single-frame trajectories as the spurious outcomes due to the aforementioned limitations of the method. This strong assumption needs to be handled cautiously since it means those vortex trajectories with their lifetimes shorter than $\Delta t_s^+ \approx 1.5$ are categorised to be unphysical. In fact, it is known that those buffer-layer small-scale vortices have rather short lifetimes corresponding to their sizes that can be affected by the temporal resolution of the current dataset (del Álamo et al. (2006), Lozano-Durán and Jiménez (2014)). However, it should be also noted that the surface-area filtering operation performed prior to the tracking works in a favourable direction to the current efficiency assumption, by eliminating those microscopic vortices with very short lifetimes.

In order to characterise those vortex trajectories with their length equals to unity (i.e. singular events), the probability density function of the volumes of the point trajectories is plotted in Figure 5.9(a) with a red solid line (—), and compared with the volumes of the rest of the vortex structures belong to the trajectories whose length is more than unity (– –). It can be seen that the volume distribution of the probability of the single-frame trajectories are indeed constraint towards the smaller end of the volume spectrum in comparison to the corresponding quantity of the longer trajectories. Note, however, there is still significant probability of the single-frame trajectories consist of the vortices with, for instance, $\approx 10\%$ of the numerical box volume which should have much longer lifetimes than Δt_s. Those single-frame trajectories with the relatively high-volume vortices can be therefore considered as the spurious outcomes due to the limitations of the current tracking method.

Figure 5.9(b) shows the corresponding probability function of the centroid locations of the single-frame trajectories. It can be seen that the high probabilities are preferentially located above the three no-slip boundaries, with the highest peak located above the bottom-wall bisector. Note the global probabilistic peak location coincides well with the location of the maximum averaged volume presented in Figure 5.5.

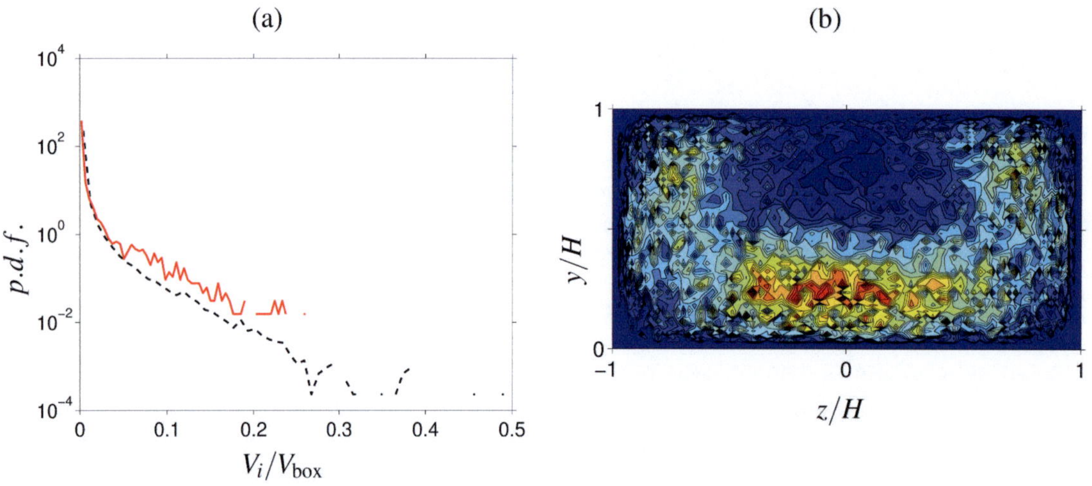

Figure 5.9: (a) Probability density function of the tracked vortex volume normalised by the numerical box volume V_{box}. Different coloured lines indicate: red solid, vortex connections with single vortex; black dashed, vortex connections with more than one vortices. (b) Probability density function of the centroid locations of the vortex trajectories with the length equals to unity.

Consequently, let us now temporarily suppose the high probability of the existence of the spurious single-frame trajectories are associated with the large-scale vortex structures preferentially located in the region. It is then possible to construct a plausible scenario that the appearances of the spurious single-frame trajectories are consequences of the vortex merging events, based on the fact that those large-scale structures have higher chance of collisions due to the larger occupied space, and should be advected at a velocity different from the advection velocities of the neighbouring small-scale structures (cf. del Álamo and Jiménez (2009)). At this point, please recall the aforementioned implementation feature of the current method that every vortices are marked as initial points or successors of trajectories strictly once. It is then possible to imagine some cases that some arbitrary small-scale structure which in fact merges into its neighbouring large-scale structure cannot mark the structure as its successor and terminates the tracking attempt artificially at that point of time. It is simply because the large-scale structure has been already marked as a part of the other trajectories that merges with the structure during the same time interval, and traced prior to the considered trajectory due to the iterative feature of the tracking procedure. This plausible scenario therefore motivates once again the improvements of the current method by incorporating a graph-based connectivity management system as discussed earlier.
In order to minimize the contributions from such spurious trajectories with premature terminations to the upcoming results, we shall eliminate those single-frame trajectories from the forthcoming trajectory statistics.

5.4.2.1 Lifetimes and ages of vortices

At last, we have reached to this point where the life of the typical open duct vortices can be discussed. For this part of the statistical analysis of the vortex trajectories, it is worth reminding ourselves that, unlike the canonical plane channel flow, our duct flow has two inhomogeneous directions that require a much larger number of statistical samples to be converged satisfactorily. It is

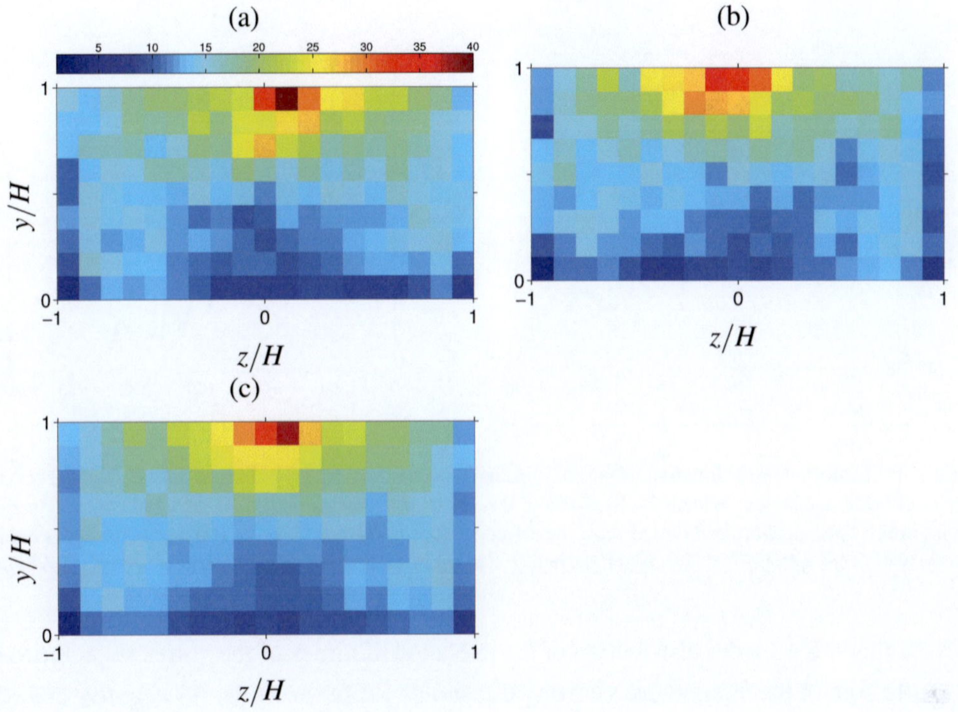

Figure 5.10: Spatial distribution of averaged vortex lifetime $\langle \tilde{T}_l \rangle$ normalised by viscous time unit ν/u_τ^2. (a) Vortex with *positive* ω_x; (b) *negative* ω_x; (c) average of (a) and (b). The vortex lifetime was associated with every terminal centroid locations of the tracked vortices. The statistics were then averaged over square boxes enclosed by 10-by-20 equidistant grid points (not CGL points).

therefore particularly challenging for the statistical analysis of the vortex trajectories, since some of the statistical quantities, such as the averaged lifetimes of the vortex trajectories as function of the locations on the duct cross-section, can be accumulated only once per trajectories as one-dimensional pointwise data on the two-dimensional cross-section. By considering the fact that the number of connected vortices averaged over all trajectories is approximately 10.5 ($\approx 15.8\nu/u_\tau^2$), such kind of per-trajectory statistics reduces the number of data sample drastically by the equivalent factor on average. To compensate such severe requirements, we shall therefore employ much coarser two-dimensional averaging grid (such as the one in Figure 5.5) for improving the statistical convergence throughout this part of analysis. The grid spacing was designed to be equidistant in both free-surface normal and spanwise directions with approximately 17 and 16 viscous length scales respectively.

First, the spatial distribution of the averaged vortex lifetimes was investigated with a special emphasis upon the dynamics of the vortices near the free-slip plane. We differentiate the age T_l and lifetime $\tilde{T}_l$ of every vortex trajectories l by defining as follows:

$$T_l : [t_l^0, t_l^{N_l}] \to [0, \tilde{T}_l] \tag{5.6}$$

where $T_l(t) = t - t_l^0$ whereas N_l is the number of connected vortices in the trajectory l. In other words, every vortex trajectories l have individual life histories from the birth (i.e. the initial point of the trajectory) with the age equals to zero, through the intermediate phase and until the death (the endpoint of the trajectory) where the age reaches to the lifetime $\tilde{T}_l$.

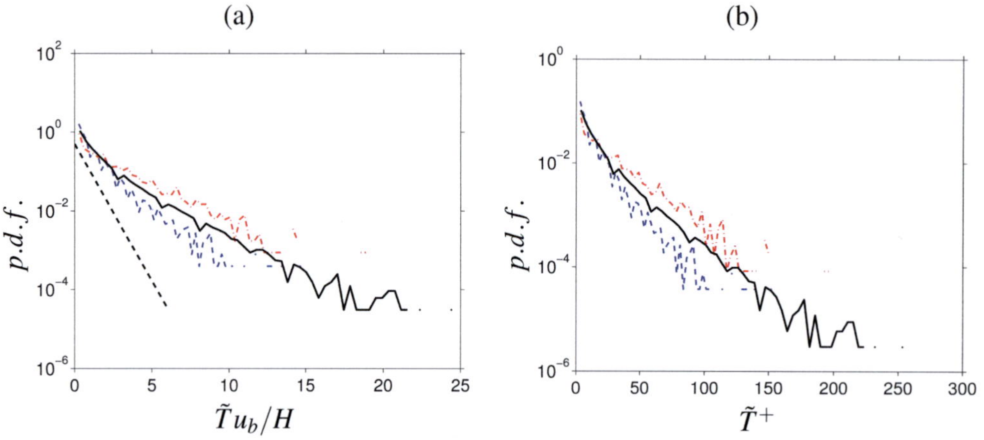

Figure 5.11: Probability density function of vortex lifetime $\tilde{T}$ normalised by (a) bulk time unit H/u_b; (b) viscous time unit ν/u_τ^2. Different line styles indicate the lifetimes of the vortex trajectories with their terminal centroid locations in: black solid, entire cross-section; red dotted-dash, near free-surface ($y/H > 0.7$, $y^+ > 105$); blue dashed, near bottom-wall no-slip wall ($y/H < 0.3$, $y^+ < 45$). Black dashed line in (a) is a reference value indicates $\sim \exp(-1.6\tilde{T} u_b/H)$.

Figure 5.10 shows the spatial distribution of the averaged lifetimes of the vortex trajectories conditioned by the sign of the streamwise vorticity thresholds as: (a) *positive*, (b) *negative* and (c) both vorticity values. It is important to notice that the global peaks of all three plots are consistently located at the duct mid-span directly below the free-slip plane. The observed results are partially expected since the peak region is the most distant location from any of the no-slip walls with the minimum shear, which is known to have the most significant effect on the vortex lifetimes to be short. However, it is still interesting to observe that the presence the impermeable boundary condition has seemingly no negative influence on the vortex lifetimes, despite the fact that the vortices are highly populated along the free-slip plane due to the boundary condition and therefore higher probability of vortex-vortex interactions in the region. Furthermore, it is important to mention that those long-living near free-slip plane vortices are relatively small, which can be deduce from the lifetime peak location being very close to the free-slip plane. This particular feature of the observed results is quite different from our previous knowledge based on the studies on the vortex dynamics in the canonical flow, which generally associate the longer vortex lifetimes with their larger sizes (cf. Lozano-Durán and Jiménez (2014)). Note also the observed results are consistent with our vortex-sorting hypothesis that the streamwise vortices with certain sense of rotation generated on the side-wall regions are transported towards the duct mid-span and end their lives in the region.

Probability density functions of the vortex lifetimes conditioned by the mean centroid location of the vortex trajectories in free-slip plane normal direction are also shown in Figure 5.11. It is visible that the vortex trajectories directly beneath the free-slip plane have significantly higher probability of survival for longer periods of time compared to the structures that are directly above the bottom-wall no-slip boundary. Note also the position-independent initial exponential decay rate of approximately -1.6 reported by Lehew et al. (2013) from their boundary-layer experiments is present in the current results (cf. Figure 5.11(a)).

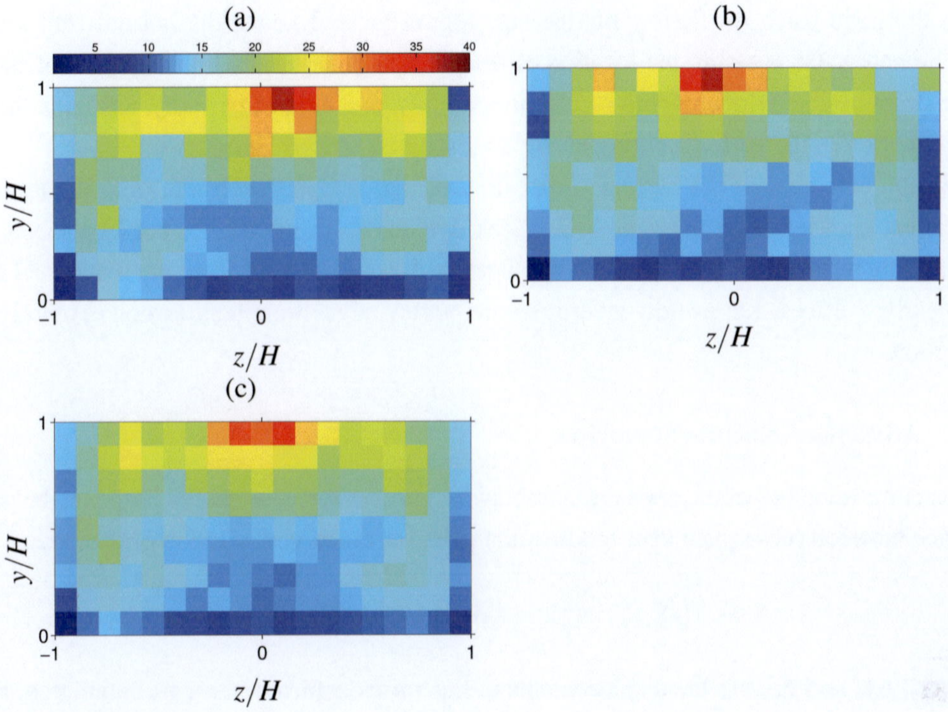

Figure 5.12: Spatial distribution of averaged vortex age $\langle T_l \rangle$ normalised by viscous time unit ν/u_τ^2. (a) Vortex with *positive* ω_x; (b) *negative* ω_x; (c) average of (a) and (b). The vortex age was associated with every centroid locations of the tracked vortices, therefore intermediate contributions along the trajectories are included. The statistics were then averaged over square boxes enclosed by 10-by-20 equidistant grid points (not CGL points).

The averaged vortex age distribution contains additional information about the intermediate history, and presented in Figure 5.12 with the identical categorisation as in Figure 5.10. It can be seen that the regions of the high-average age are spread along the free-slip plane whilst the peak locations are slightly shifted towards either side of the side-walls depending on the sign of rotation of the vortices (cf. Figure 5.12(a,b)). The first observation complies with our assumption that the vortices near the free-slip plane are predominantly transported from the side-walls towards the duct-mid span, ageing gradually during their journeys. The second observation on the rotational direction dependency of the age peak locations additionally supports our vortex-sorting hypothesis. In other words, those streamwise vortices rotating in clockwise travel predominantly from a location near the side-wall at $z/H = -1$ towards the duct mid-span by interacting with the free-slip plane due to the mirrored dipole effect, whereas the vortices rotating in the counter-clockwise direction travel predominantly from the opposite-side side-wall toward the centre.

On the other hand, relatively young vortices are distributed within the mixed-boundary corners independent of the rotational directions of the streamwise vortices. Such trend again follows our expectation based on the fact that the mean side-wall shear stress distribution marks its peak in the regions, destroying the mixed-corner vortices in a quick manner. A pair of tails of the high vortex-age regions visible in Figure 5.12(c), which are extending from the side-wall regions and circumventing the mixed-boundary corners, are also interesting to observe. It possibly implies that probable routes of the side-wall vortices exist around the side-wall bisectors, advected towards the free-slip plane by the motions corresponding to the mean outer secondary flow.

Finally, the large patch of the low-lifetime/age region located above the bottom-wall mid-span should be noted. Once again, the location of the patch is consistent with the region of the high-volume structures, and the observed results are possibly, and at least partially, artificial due to the high probability of the vortex merging events as discussed earlier.
Over all, the outcomes of the current vortex lifetime/age analysis seem reasonable compared to the previous knowledge on the near-wall vortex dynamics. Furthermore, our hypothesis of the vortex-sorting mechanism near the free-slip plane gains some support from the dynamical aspects of the nearby vortices. Let us now investigate the vortex advection phenomenon explicitly in the next section.

5.4.2.2 Advection velocities of vortices

The advection velocities of the open duct streamwise vortices were computed based on the position difference between subsequent centroid location pairs for every vortex trajectories l, viz.

$$([u_{\mathrm{v}}^m, v_{\mathrm{v}}^m, w_{\mathrm{v}}^m])_l = ([\Delta x_{\mathrm{v}}^m, \Delta y_{\mathrm{v}}^m, \Delta z_{\mathrm{v}}^m])_l / \Delta t_{\mathrm{s}} \tag{5.7}$$

where Δx_{v}, Δy_{v} and Δz_{v} are the displacements of the vortex centroid in x-, y-, z-direction, respectively, from time t^m to time t^{m+1} over the time interval $m = 1 \ldots N_l - 1$. The associated locations of the advection velocities were set at the mid-points of the each centroid pair, viz.

$$([x_{\mathrm{v}}^m, y_{\mathrm{v}}^m, z_{\mathrm{v}}^m])_l = \left(\left[\frac{x_{\mathrm{c}}^m + x_{\mathrm{c}}^{m+1}}{2}, \frac{y_{\mathrm{c}}^m + y_{\mathrm{c}}^{m+1}}{2}, \frac{z_{\mathrm{c}}^m + z_{\mathrm{c}}^{m+1}}{2} \right] \right)_l . \tag{5.8}$$

First, the spatial distribution of the averaged streamwise advection velocities was investigated and presented in Figure 5.13 with the same categorisation as in Figure 5.10 and 5.12. Additionally, Figure 5.13(d) shows the mean streamwise velocity contour $\langle u \rangle$ for comparison.
It is visible that all averaged advection velocity distributions follow the corresponding mean streamwise velocity closely, being consistent with the observations from the plane channel flow. Furthermore, the levels of the advection velocity adjacent the no-slip walls are in a good agreement with the previously reported value $\approx 11 u_\tau$ except the solid-solid corners where the levels drop significantly. Note finally that there is no visible vortex rotational-direction dependency in this quantity (cf. Figure 5.13(a,b)).
The rotational-direction independence also holds for the distributions of the free-slip plane normal advection velocity presented in Figure 5.14. There is, however, one major discrepancy located in the duct core region above the bottom-wall bisector where the global maximum, which is absent from the corresponding mean velocity field (cf. 5.14(d)), is situated. This discrepancy can be explained again from the viewpoint of the vortex merging events. The explanation is based on the fact that the large-scale vortex structures in general have higher centroid locations in y-direction than the ones of the small-scale near-wall vortices, so that the centroid locations rises suddenly in a case of the vortex merging. Apart from the exception, the advection velocity distribution is once again in a good agreement with the mean velocity field $\langle v \rangle$.
In contrast, the spatial distribution of the spanwise vortex advection velocities has a significant vortex rotational-direction dependence, both near the free-slip plane and in the bottom-half of the

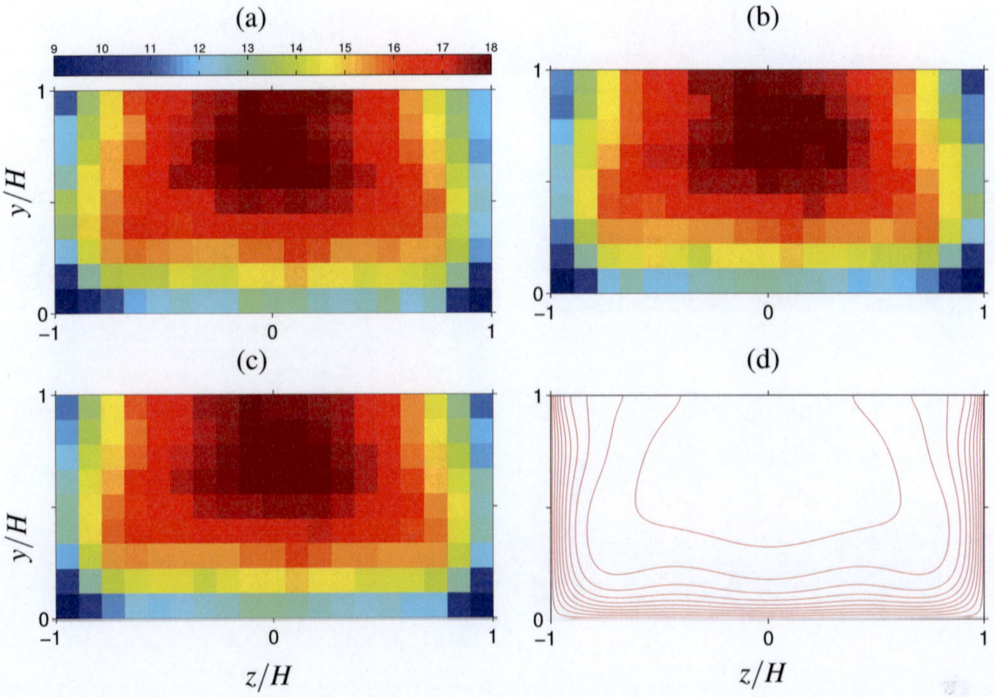

Figure 5.13: Spatial distribution of averaged vortex streamwise advection velocity $\langle u_v \rangle$, normalised by u_τ. (a) Vortex with *positive* ω_x; (b) *negative* ω_x; (c) average of (a) and (b). The advection velocity was computed from the centroid location differences of the tracked vortices in each pair of consecutive snapshots, and associated with the mid-point of the centroid locations. The statistics were then averaged over square boxes enclosed by 10-by-20 equidistant grid points (not CGL points). (d) Mean streamwise velocity contours $\langle u \rangle$ with the level of the contour lines indicating 0.1(0.1)0.9 times $\max\{\langle u \rangle\}$.

duct cross-section (cf. Figure 5.15(a,b)). It is visible that there are the band-like regions of the high advection velocities along the free-slip plane that reach slightly beyond the duct mid-span, then meeting the opposite-signed peak with a rounder shape.

The directional dependency of the spanwise advection velocities in the bottom-wall regions is also a very interesting phenomenon, which provides to obtain an additional explanation to the generation mechanism of the secondary motions in the closed duct sense.

It can be seen that the arrangement of the spanwise velocity distribution is consistent, though it is weaker in the intensity, with the vortex-sorting hypothesis if we apply the same explanation to the bottom-wall boundary as well. In this scenario, the near-bottom-wall vortices rotating in the clockwise direction move towards the left side-wall and the opposite is true for the vortices rotating in the anti-clockwise direction. At this point, it is important to recall that the dipole hypothesis employed to explain the vortex dynamics near the free-slip plane is essentially an inviscid two-dimensional process, which do not discriminate between the free-slip and no-slip boundaries since the necessary condition is solely the impermeability in the boundary-normal direction. There is, however, a crucial modification imposed by the no-slip condition that those vortices near the no-slip walls cannot approach towards the boundaries as easily as for the near free-slip plane vortices, whilst the distance to the boundary plays an important role as demonstrated in 3.3.4. Note this wall-distance factor might explain the slower spanwise movements. Nevertheless, it is still possible to conjecture a meaningful explanation with some caution by the mechanism as it was

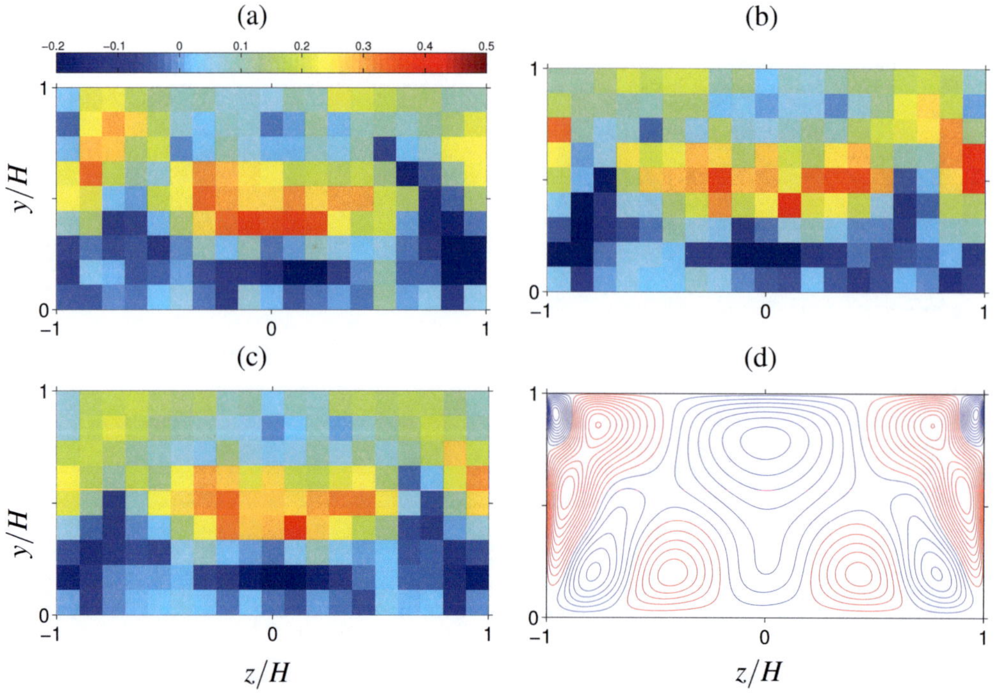

Figure 5.14: Spatial distribution of averaged vortex free-slip plane normal advection velocity $\langle v_\mathrm{v} \rangle$, normalised by u_τ. (a) Vortex with *positive* ω_x; (b) *negative* ω_x; (c) average of (a) and (b). The advection velocity was computed from the centroid location differences of the tracked vortices in each pair of consecutive snapshots, and associated with the mid-point of the centroid locations. The statistics were then averaged over square boxes enclosed by 10-by-20 equidistant grid points (not CGL points). (d) Mean free-slip plane normal velocity contours $\langle v \rangle$ where red and blue lines indicate *positive* and *negative* velocities respectively. The level of the contour lines is -0.9(0.1)0.9 times $\max\{\langle v \rangle\}$.

successfully applied by Sekimoto et al. (2011) (cf. their figure 18) though in a different context of the interactions between the buoyancy-driven large-scale vortex and the near-wall small-scale vortices in the solid-solid boundary corners.

The averaged distribution of Figure 5.15(a) and (b) shown in Figure 5.15(c) recovers the anti-symmetry about the duct mid-span as in the mean spanwise velocity field presented in Figure 5.15(d).

Figure 5.16 shows the velocity vector field based on the cross-sectional advection velocity distributions (i.e. $\langle v_\mathrm{v} \rangle$ and $\langle w_\mathrm{v} \rangle$), which deviates significantly from the mean secondary flow pattern chiefly from the discrepancy in $\langle v_\mathrm{v} \rangle$. This advection velocity pattern should be re-examined once the error-contribution from the artificial movements in y-direction will be eliminated by improving the current implementation.

Finally, we shall investigate the spatial distribution of the averaged trajectory path angles $\langle \theta_\mathrm{v} \rangle$. The trajectory path angles can be computed for every time interval m, based on the streamwise and spanwise displacements, viz.

$$(\theta_\mathrm{v}^m)_l = (\tan^{-1}(\Delta z_\mathrm{v}^m / \Delta x_\mathrm{v}^m))_l \, , \tag{5.9}$$

where $-\pi \leq \theta_\mathrm{v} \leq \pi$, and the trajectory path angles are associated with the lth structure's centroid mid-points as before. Note a positive path angle indicates movements in the positive z-direction, whereas a negative angle indicates movements in the negative z-direction.

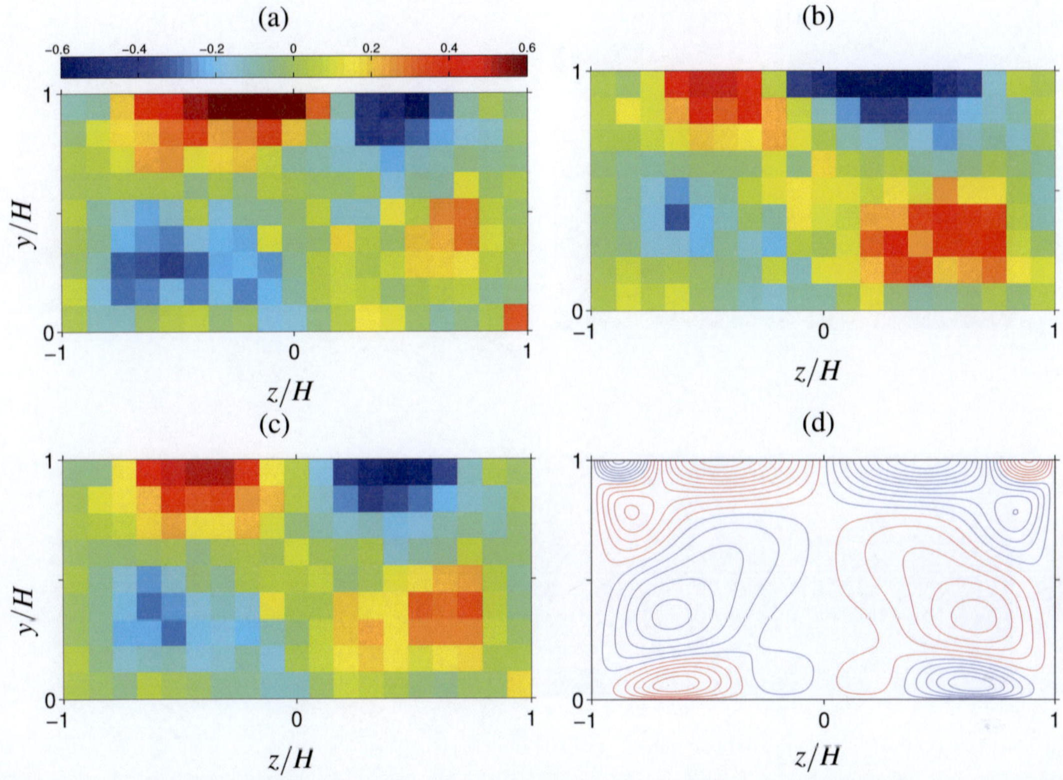

Figure 5.15: Spatial distribution of averaged vortex spanwise advection velocity $\langle w_\mathrm{v} \rangle$, normalised by u_τ. (a) Vortex with *positive* ω_x; (b) *negative* ω_x; (c) average of (a) and (b). The advection velocity was computed from the centroid location differences of the tracked vortices in each pair of consecutive snapshots, and associated with the mid-point of the centroid locations. The statistics were then averaged over square boxes enclosed by 10-by-20 equidistant grid points (not CGL points). (d) Mean spanwise velocity contours $\langle w \rangle$ where red and blue lines indicate *positive* and *negative* velocities respectively. The level of the contour lines is -0.9(0.1)0.9 times $\max\{\langle w \rangle\}$.

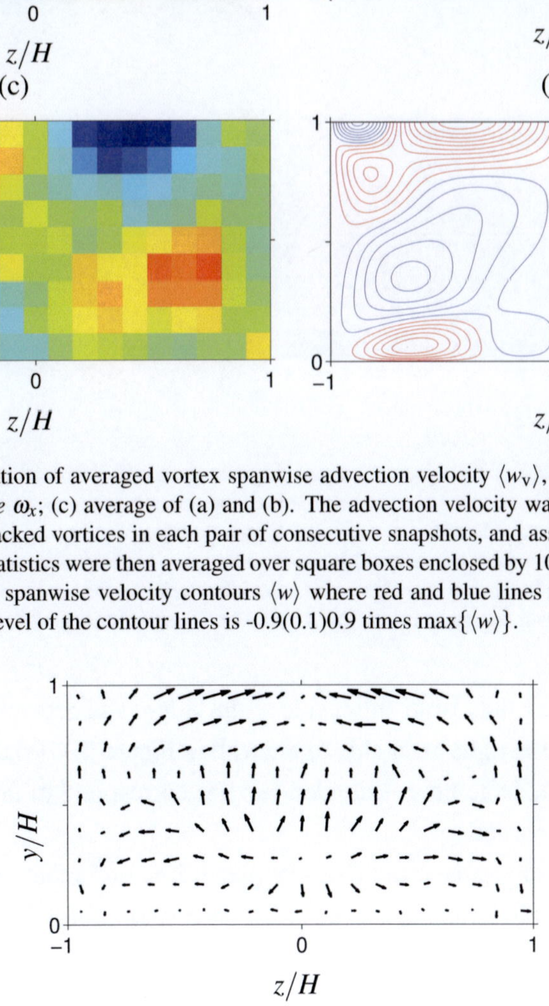

Figure 5.16: Mean cross-plane vortex advection velocity vectors, computed from $\langle v_\mathrm{v} \rangle$ and $\langle w_\mathrm{v} \rangle$.

The corresponding spatial distributions are presented in Figure 5.17. Firstly, note that the magnitude of this quantity remains very small so that even the largest end of the colour-scale is as small as $\approx \pi/52$ radian ($\approx 3.5°$), which is due to the much smaller cross-sectional advection velocities in comparison with the streamwise advection. Secondly, the regions of relatively large path angles exist along the side-walls including the mixed-boundary corners, with the sign of the angles indicating the trajectories travel towards the side-wall regions on average. Furthermore, there is a noticeable rotational-direction dependency exists in the pattern of the side-wall peak that forms a

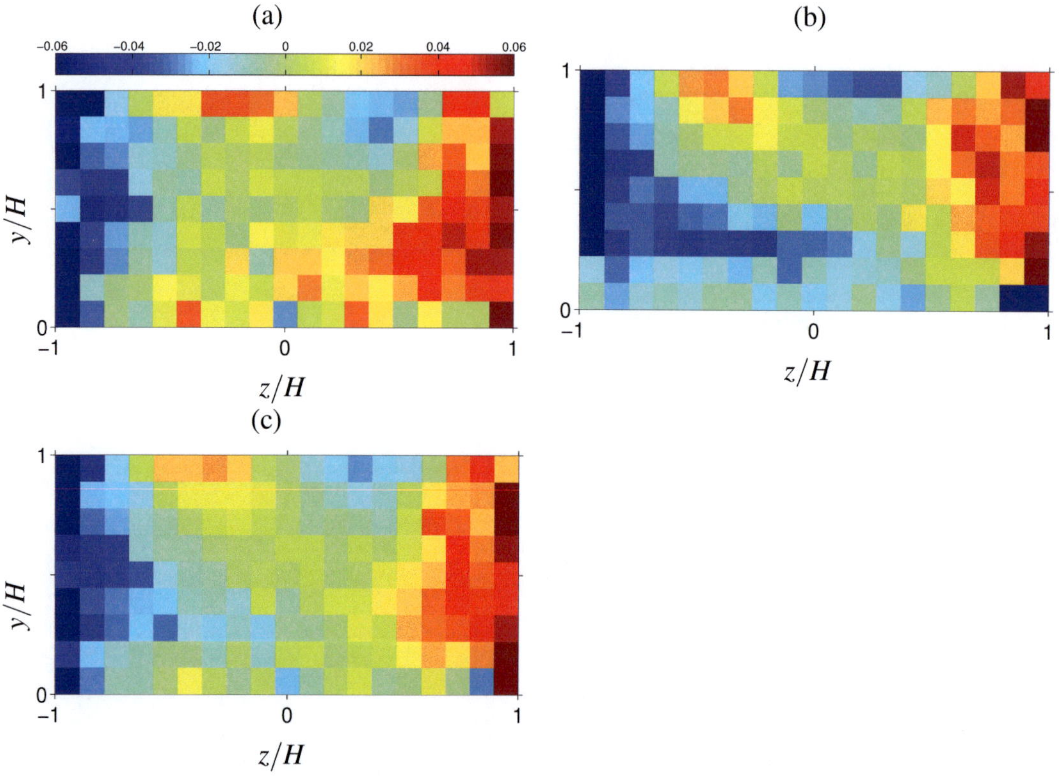

Figure 5.17: Spatial distribution of averaged vortex path angle in radian $\langle\theta_v\rangle$ defined on xz-plane, such that *positive* angles indicate spanwise movement towards right-hand-side of the domain, whereas *negative* angles indicate opposite movements towards the left-hand-side of the domain. (a) Vortex with *positive* ω_x; (b) *negative* ω_x; (c) average of (a) and (b). Each θ_v was computed from the centroid location differences in streamwise and spanwise directions of the tracked vortex pair from the consecutive snapshots, and associated with the mid-point of the centroid locations. The statistics were then averaged over square boxes enclosed by 10-by-20 equidistant grid points (not CGL points).

triangular profile extending only from either side of the side-walls depending on the sense of rotation (cf. the red patch on the right-hand-side of the wall in Figure 5.17(a) and the blue patch on the opposite side-wall in (b)). Note those extended patches correspond to the bottom-wall spanwise velocity peaks discussed earlier.

Finally for the trajectory angles near the free-slip plane, it is important to notice the existence of the two regions of relatively large angles, which is once again consistent with the vortex dipole hypothesis being associated with the generation of the outer secondary motions. Moreover, the small magnitudes of those trajectory angles explain the difficulty to detect this kind of vortex transporting phenomenon both numerically and experimentally.

5.5 Conclusion

The instantaneous three-dimensional streamwise vortex structures of an open duct flow with $Re_b = 2205$ and $A = 1$ were extracted, and their spatial distributions and dynamics were analysed. The investigations of the vortex dynamics were realised by tracking the extracted vortices in time by a new method developed for this study. The proposed temporal-tracking method is based on the centroid locations of the vortices and their overlapping volumes between the subsequent snapshots.

First by examining the extracted vortices independently, it was confirmed that those structures indeed carry the majority of the streamwise vorticity responsible for the manifestation of the open duct mean streamwise vorticity pattern investigated in the previous chapters. It was also found that the preferential locations of the centroid locations of the extracted vortices reconstruct qualitatively the mean streamwise vorticity pattern, that is consistent with the results from the previous coherent structure analysis based on the locations of the vortex cores identified on each duct cross-section. Moreover, the existence of the large-scale streamwise vortices was identified, that are preferentially located above the bottom-wall bisector. The possible consequences of those large-scale vortices to the accuracy of the proposed tracking method was discussed, and some suggestions for the further improvement of the current method were made.

Subsequently, the lifetime statistics of the tracked vortex trajectories were examined, and the high concentration of the small-scale vortices with the longest lifetime was identified directly below the free-slip plane bisector. The observed result was confirmed to be consistent with the vortex-sorting hypothesis that was proposed to explain the generation mechanism of the open duct secondary motions. Note the hypothesis is based on the interactions between the small-scale near-wall vortices and the free-slip boundary, supposed to be only effective in the direct vicinity of the boundary.

The cross-sectional distributions of the advection velocities were analysed separately in each direction. It was found that the vortex advection velocity patterns follow the mean primary and secondary velocity patterns closely as in the plane channel flow, except a part of the distribution of the free-slip plane normal velocity component. An explanation for such discrepancy was attempted by associating it with the possible tracking inaccuracy caused by the merging events with the aforementioned large-scale vortices.

Most importantly, the vortex rotational direction dependency was found in the spanwise advection velocity distribution both directly below the free-slip plane and the area closer to the bottom-wall. Such observations are again consistent with the vortex-sorting hypothesis, providing an additional support for the proposed generation mechanism of the secondary motions in the turbulent duct flow.

Finally the statistics of the trajectory angles projected onto the plane parallel to the bottom-wall boundary was studied. Relatively strong tendency of the vortices moving towards the duct sidewalls were detected, together with significantly weaker but still noticeable distributions of the path angles below the free-slip plane supporting the vortex-sorting hypothesis. The latter small averaged trajectory angles underline the difficulty of identifying such motions below the free-slip plane, indicating a possible reason why such phenomenon have not caught much attention previously, despite the long history of the research on this topic.

To conclude, it is important to note that all statistical results of the open duct vortex dynamics presented in this chapter support consistently the vortex-sorting hypothesis proposed in Chapter 3. Although our current standpoint is still far from being conclusive to prove the causality of the proposed mechanism to the open duct secondary motions, it was still crucial to demonstrate the possibility based on the actual three-dimensional turbulent flow fields. A series of such attempts allow us to progress beyond the conceptual modelling carried out in this thesis, towards our ultimate goal to demonstrate the causality.

Chapter 6

Conclusion

6.1 Summary

Mean secondary motions in the open duct flow with rectangular cross-section and infinite streamwise extension were studied by means of fully-resolved pseudo-spectral direct numerical simulations. The rigid free-slip plane approximation was employed to represent the free surface, which implies that the flow are in the zero Froude and Weber number limit. In total 38 open duct, and 10 additional closed duct simulations were performed for this study, with Reynolds number varied from the marginal to the fully-developed regime up to $Re_b = 7000$, over a range of aspect ratio $0.5 \leq A \leq 8$. The performed simulations are summarised in terms of those two relevant parameters in Figure 6.1, namely bulk Reynolds number and aspect ratio, and presented together with the parameters of the studies performed prior to the current investigation.
The above parameter values were chosen specifically to fulfil the following knowledge gaps in the study of the open duct turbulence:

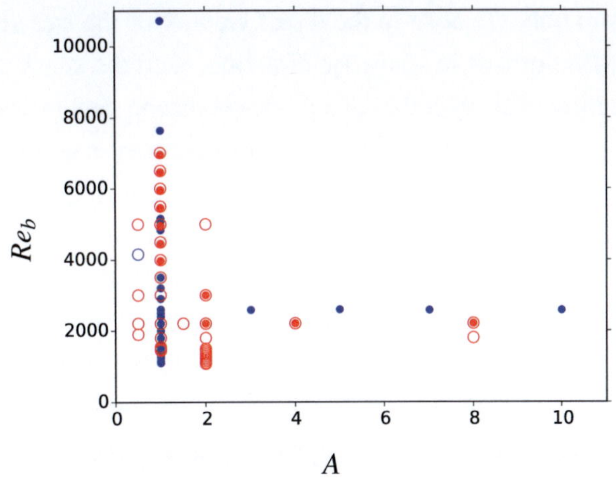

Figure 6.1: Parameter map of the duct direct numerical simulations. The red symbols represent the simulations performed for the current study, while the blue symbols represent the previous studies by: Gavrilakis (1992), Huser and Biringen (1992), Uhlmann et al. (2007), Pinelli et al. (2010), Raiesi et al. (2011) and Vinuesa et al. (2014) for closed duct; Joung and Choi (2010) for open duct. Open and filled symbols represent open and closed duct configurations respectively.

1) key coherent structures and their dynamics responsible for the open duct secondary motions;
2) existence of any clear marginal flow features, such as the four-vortex state observed in the closed duct flow;
3) Reynolds number and aspect ratio dependence in the mean secondary flow patterns in the fully-developed turbulence regime;
4) Reynolds number and aspect ratio dependence in the dip-phenomenon;
5) both qualitative and quantitative similarity and difference with respect to the closed duct secondary motions.

In the following, our findings regarding those knowledge gaps will be discussed individually, except for the aspect 5) that will be mentioned throughout.

Key coherent structures and their dynamics

Through an extensive vortex eduction analysis, the quasi-streamwise vortices preferentially located in the mixed-boundary corner region, rotating towards those corners, were identified as the key coherent structures responsible for the formation of the mean inner secondary flow. Moreover, the quasi-streamwise vortices existing directly below the free-slip plane with the opposite sense of rotation to the inner secondary vortices were found to contribute to the mean outer secondary motions. The probability of the existence of the mixed-boundary corner quasi-streamwise vortices was found to be predominantly high in comparison with the rest of the duct cross-section, with their probable location tightly concentrated in the mixed-boundary corner regions. In contrast, the locational preference of the vortices associated with the outer secondary flow is distributed broadly along the free-slip plane.

Such unique probabilistic distribution of the vortices was conjectured as a consequence of the vortex sorting mechanism only effective in the direct vicinity of the free slip-plane, which transports the quasi-streamwise vortices in spanwise direction, with the sense of direction depending on their direction of rotation. This hypothesised transport mechanism is based on the study of the two-dimensional vortices interacting with the different boundary conditions by Orlandi (1990). The key idea is that a two-dimensional vortex located close to a free-slip boundary sees its own mirrored image due to the imposed symmetric boundary condition, forming a vortex dipole with itself that induces the movement of the vortex in the direction parallel to the free-slip boundary.

In order to demonstrate the existence and its significance of this essentially inviscid two-dimensional process in the three-dimensional open duct turbulence, the instantaneous three-dimensional streamwise vortex structures of an open duct flow with $Re_b = 2205$ and $A = 1$ were extracted, and their spatial distributions and dynamics were analysed. The investigations of the vortex dynamics were realised by tracking the extracted vortices in time by a new method developed for this study.

Within the study of the open duct vortex dynamics, the lifetime statistics of the tracked vortex trajectories were examined, and the high concentration of the small-scale vortices with the longest lifetime was identified directly below the free-slip plane bisector. The study also demonstrated the existence of the vortex rotational direction dependence in the spanwise vortex advection velocity distribution both directly below the free-slip plane and the area close to the duct bottom-wall. Moreover, the statistics of the vortex trajectory angles projected onto the plane parallel to the free-slip plane showed the distribution of the path angles away from or towards the duct side-walls depending on the rotational direction of vortices existing below the free-slip plane. All above observations hold a consistency with the aforementioned vortex-sorting hypothesis.

Marginal open duct turbulence with various aspect ratios

It was found that the minimal sustainable Reynolds number limits of the open duct turbulence are function of aspect ratio as in the closed duct counterpart, which were determined for three aspect ratios within this study, namely: $Re_b = 1900$ ($Re_\tau \approx 135$) for $A = 0.5$, $Re_b = 1450$ ($Re_\tau \approx 102$) for $A = 1$ and $Re_b = 1080$ ($Re_\tau \approx 74$) for $A = 2$.

Unlike the marginal open duct turbulence with $A = 1$, the flow with $A = 0.5$ and $A = 2$ exhibit drastic changes with their mean secondary flow patterns when the flow Reynolds number is sufficiently close to the aforementioned minimal limits. A unique four-vortex secondary flow pattern, with only the duct side-walls hosting one low-velocity streak each, appears in the open duct flow with $A = 0.5$ instead of the standard six-vortex pattern observed in the higher Reynolds number flow. Similarly, a two-vortex secondary flow pattern with only one low-velocity streak statistically located at the bottom-wall boundary bisector was observed in the marginal open duct flow with $A = 2$.

In both marginal flow configurations, the mechanism of the disappearance of the velocity streaks from the no-slip walls was shown as a consequence of the boundary edge lengths in wall units being too short to host a minimum set of the near-wall turbulent structures (i.e. one low-velocity streak franked by a pair of counter-rotating quasi-streamwise vortices). In such scenario, the turbulent structures cannot be sustained near those restrictive walls, leading to the turbulent structures to be detached from those no-slip walls, similar to the phenomenon observed in the closed duct transitional turbulence with large aspect ratios (cf. Takeishi et al. (2015)).

In cases of the marginal turbulence with $A = 0.5$ and $A = 1$, it was shown that those flow do not exhibit any temporal-alternation of orientation observed in the marginal square closed duct flow of Uhlmann et al. (2007). On the other hand, the marginal flow with $A = 2$ indicated the existence some kind of temporal-alternation of orientation, which requires further investigations.

Reynolds number dependence of fully-developed duct turbulence

Considering the mean secondary flow pattern, the size and location of the inner secondary mean vortex exhibit strong Reynolds number dependence, becoming continuously smaller and moving towards the mixed-boundary corners as Reynolds number increases. In other words, the corresponding streamfunction extrema maintain approximately constant distances to the free-/no-slip boundaries in wall units. Such inner-scaling behaviour of the secondary flow pattern is unique to the open duct flow, since the secondary flow pattern in the closed duct counterpart is known to be Reynolds number independent (cf. Pinelli et al. (2010)). The aforementioned vortex eduction study revealed that the inner-scaling property of the mean inner secondary flow is due to the strong

link between the mean velocity field and the near-wall quasi-streamwise vortices, whose existence was shown to be highly probable in the mixed-boundary corner regions. The rest of the mean secondary flow features, such as the locations and dimensions of the outer secondary flow vortices and the bottom-wall mean vortices residing next to the no-slip boundary corners, were found to be Reynolds number independent, i.e. outer-scaling.
The mean streamwise vorticity pattern also exhibits a unique scaling phenomenon in case of the open duct flow, namely the outer-scaling of the size and location of the outer secondary vorticity cells, which is only observed in spanwise direction. This unique phenomenon was explained in terms of the vortex-sorting hypothesis, in which the quasi-streamwise vortices contributing to the outer secondary vorticity cells are transported along the free-slip plane. The limit of such spanwise vortex movements imposed by the duct geometry was associated with the outer-scaling behaviour of the mean outer secondary vorticity cell dimension. The remaining geometrical features of the mean streamwise vorticity pattern were found to be in inner-scaling, similar to the closed duct counterparts.
The scaling of the cross-sectional averaged mean secondary flow intensity was also studied, and it was found that whilst the averaged intensity in the open duct flow scales with the bulk velocity (u_b), the same quantity in the closed duct flow was found to scale with a mixed velocity scale ($\sqrt{u_\tau^3/u_b}$), where u_τ is the friction velocity based on the wall stress averaged over all no-slip boundaries.

Aspect ratio dependence in fully-developed duct turbulence

The earlier experimental and numerical observations that the duct secondary flow are confined in the vicinity of the side-walls even in large aspect ratio ducts, were confirmed in a various aspects, including: the near-wall velocity streak locations; the secondary flow streamfunction and streamwise vorticity extrema locations; and the depth-averaged secondary flow intensity distribution in spanwise direction.
Through the examination of the depth-averaged secondary flow intensity distribution, it was also found that the averaged intensity becomes independent of aspect ratio, if we further integrate the averaged intensity in spanwise direction only over certain distance from the duct side-walls, namely the duct full-height H in the open duct, and the duct half-height h in the closed duct flow. Moreover, it was found for both open and closed duct flow, that the secondary flow intensity decays exponentially as the distance from the nearest side-wall increases, approaching asymptotically to some constant decay rate as aspect ratio increases. For the range of aspect ratios considered here, the decay constant for the open duct flow with $A = 8$ was found at $\lambda \approx 0.5$, whereas the constant for the closed duct flow with $A = 8$ was found at $\lambda \approx 0.7$.

Reynolds number and aspect ratio dependence in dip-phenomenon

The velocity dip-phenomenon is characterised by the location of the streamwise velocity maxima relocated from the free-slip plane to somewhat below, observed under certain conditions once the flow becomes turbulent. We distinguished a state of the dip-phenomenon observed over all

spanwise locations (referred as "*global dip-phenomenon*"), from the state that the velocity maxima relocation is only observed locally near the duct side-walls (i.e. "*local dip-phenomenon*").
Outer-scaling of the submerged distance from the free-slip plane at the duct mid-span under the global dip-phenomenon condition was confirmed in the open duct flow with fixed aspect ratio at unity. Furthermore, the previous hypothesis of the phenomenon to be a consequence of large-scale side-wall low-velocity streak ejection events was confirmed by means of a coherent structure eduction technique which takes into account the individual size of the educed coherent structures. It was found that the occurrence of mean dip-phenomenon can be explained by the existence of large-scale low-velocity streaks in the duct mid-span region directly below the free surface (in a statistical sense), and the corresponding submerged distance was found to be associated with the cross-sectional dimension of those streaks.
The existence of the Reynolds number dependence in the critical aspect ratio to observe the global dip-phenomenon was reported for the first time to our knowledge. Our results demonstrated that in case of low Reynolds number flow, the dip-phenomenon disappears even if the duct aspect ratio is smaller than the widely-accepted critical value of $A_{\text{crit}} \approx 2.5$. At the time of writing, the observed Reynolds number dependence in the dip-phenomenon is conjectured as a consequence of the reach of the aforementioned large-scale low-velocity streaks, extending from the duct side-walls in an oblique band-like shape, observed exclusively in the region directly below the free-slip plane.

6.2 Future perspective

Over all, the outcomes of the current research project were fruitful by successfully answering the identified key questions in the field of open duct turbulence. We expect that some of our findings are readily applicable, for instance, to improve the existing empirical models to approximate the open duct turbulence, as well as for designing the experimental studies more effectively in the topic of concern. Nevertheless, there are some open questions remaining for the future work, which can be summarised as follows.
In Chapter 3, we simulated the open and closed duct flow with $A = 1$ up to $Re_b = 7000$ ($Re_\tau \approx 400$), where the Reynolds number is approaching to the lower limit of the logarithmic layer development with its full extent. Whilst it is desirable to increase the Reynolds number limit even higher to achieve the complete development of the logarithmic layer, it is important to note that there already exists some degree of scale-separation of the turbulent fluid motions in the current flow configurations. In such cases, some special care should be taken to study the different dynamics depending on the scales of motion. Consequently, the large-scale structures in the open duct flow were studied by means of the velocity streak eduction technique which takes into account the cross-sectional dimension of the individual structures. Such kind of analysis should be extended to the large-scale circulating motions, for instance, by means of the eddy eduction technique employed for the closed square duct flow in the aforementioned MSc. thesis of Nakatsuji (2012) (*in Japanese*). Their method is based on the Clebsch potential, which can be identified as one of the two scalar functions expressing the divergence-free velocity field in a poloidal-toroidal decomposition (i.e. the function ψ in the decomposition $\boldsymbol{u} = \nabla \times \nabla \phi \boldsymbol{x} + \nabla \times \psi \boldsymbol{x}$), which can be computed from

the streamwise vorticity in each cross-sectional plane (y,z) by means of a two-dimensional Poisson problem. Similar to the large-scale velocity streak analysis, their Clebsch-potential method is capable of taking into account the individual cross-sectional scale of the educed eddy structures. Such technique should be applied to the open duct flow, and the resulting structures should be compared with the corresponding structures in the closed duct flow. Furthermore, additional comparison with the large-scale structures in the plane channel flow is also of our interest, for instance with the wall-attached vortex clusters discussed in del Álamo et al. (2006).

The distinct large-scale velocity streak structures observed just below the free-slip plane require further investigations. Their oblique-band-like pattern shown in Chapter 4 indicates the need of three-dimensional structural analysis, in addition to the cross-sectional geometrical study mentioned above. Moreover the surrounding vortical structures should also be considered, in order to obtain the complete understanding of the structural dynamics in the direct vicinity of the free-slip plane. Consequently, the obtained information should be compared with the corresponding structures in the open channel flow. Such comparative study will provide us further insight into the influence of the no-slip side-walls co-existing with the free-slip plane.

The aspect ratio dependence of the fully-developed duct turbulence discussed in Chapter 4 requires further investigation with increased Reynolds number values. In particular, the influence of the aforementioned large-scale structure to the aspect ratio dependence observed in the moderate Reynolds number flow should be addressed.

In relation to the importance of simulating the duct flow at larger Reynolds number and/or larger aspect ratio values, it is not feasible to utilise the current pseudo-spectral method. This is mainly due to the limited flexibility of the Chebyshev-Gauss-Lobatto point distribution employed in the spatial discretization in the cross-sectional directions. Such inflexibility results in extremely fine spatial resolutions in the near-wall regions, in order to maintain sufficient resolutions in the core flow regions where the point distribution is the coarsest. To overcome the difficulty, a major code upgrade which uses spectral elements only in the cross-sectional directions is under way.

The problem of sediment transport in duct flow is of particular interests of the civil engineering application. Such problem can be realised either by the point-particle approximation or the interface-resolving direct numerical simulation technique. The latter approach is particularly interesting, from the viewpoint of the dynamics of the secondary motions under the influence of the movable roughness elements.

Finally, it is highly desirable to determine the critical aspect ratio values and the critical streamwise extensions of the different marginal flow features discussed in Chapter 3 and 4. Such minimal flow units are expected to underline a kind of regeneration cycle which happens in the open duct flow. Studying such regeneration cycle is essential for the fundamental understanding of wall-bounded turbulence. In the similar context, those newly-discovered marginal mean flow patterns could potentially serve as promising initial conditions to search for open duct non-linear invariant solutions that are still to be discovered.

Appendix A

Sufficiency of grid resolution and streamwise domain length in open duct DNS

Here we present the results of the grid resolution convergence tests mentioned in Chapter 3. To examine the adequateness of the above spatial resolution criteria, a series of grid convergence test was performed for the baseline open duct case with $A = 1$ and $Re_b = 2205$, by refining or coarsening the number of modes in all three directions by a factor of ≈ 1.5, which were initially set at $[M_x, M_y, M_z] = [256, 97, 193]$. Note that the default spatial resolutions expressed in the wall units are $\Delta x^+ = 14.8$ and $\max\{\Delta y^+\} \approx \max\{\Delta z^+\} = 2.46$, whereas the coarser and finer resolutions are respectively: $\Delta x^+ = 19.9$ and $\max\{\Delta y^+\} \approx \max\{\Delta z^+\} = 3.73$; $\Delta x^+ = 9.85$ and $\max\{\Delta y^+\} \approx \max\{\Delta z^+\} = 1.85$. Mean wall stress distributions were chosen to be the measure of the grid resolution sufficiency (cf. Figure A.1), due to the high sensitivity to the spatial resolution.

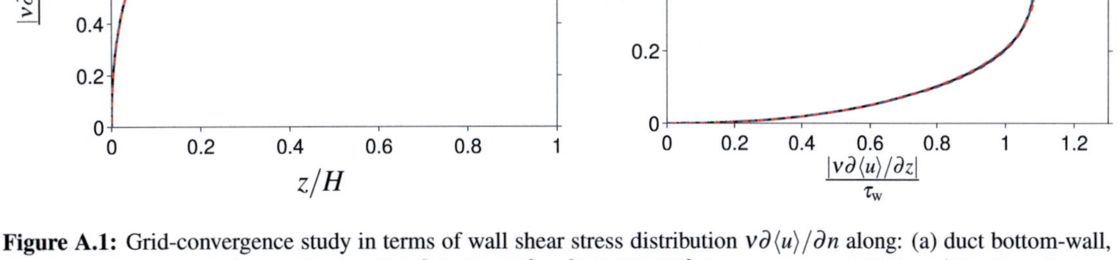

Figure A.1: Grid-convergence study in terms of wall shear stress distribution $\nu\partial\langle u\rangle/\partial n$ along: (a) duct bottom-wall, (b) side-wall. Lines indicate: black solid, $[M_x, M_y, M_z] = [256, 97, 193]$ (averaged over 12940 u_b/H); blue dashed, $[M_x, M_y, M_z] = [192, 65, 129]$ (4580 u_b/H); red dotted dash, $[M_x, M_y, M_z] = [384, 129, 257]$ (2674 u_b/H) respectively.

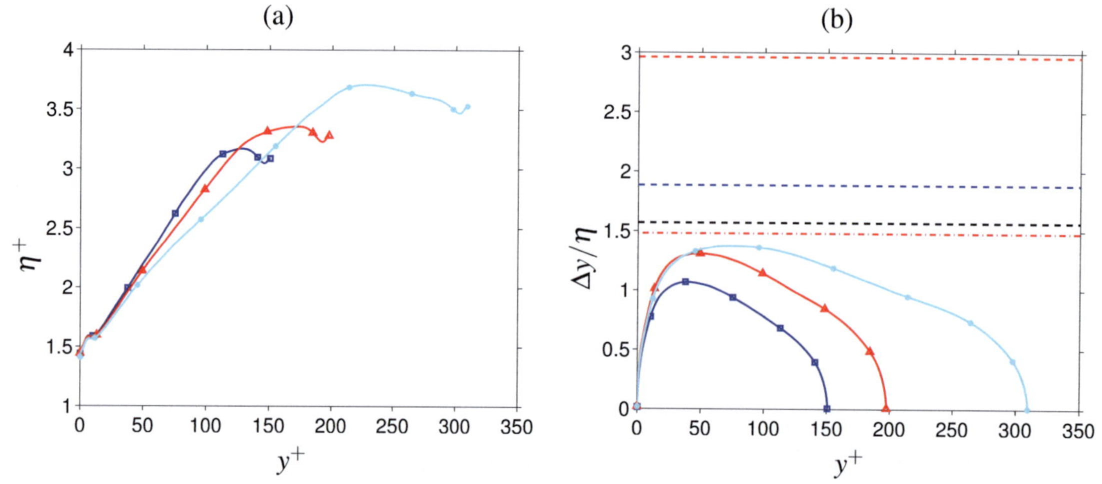

Figure A.2: (a) Local Kolmogorov length scale $\eta = (\nu^3/\varepsilon)^{1/4}$ vs. free-slip plane normal distance y^+; (b) ratio of the wall-normal grid spacing $\Delta y(y_i)$ to the local η. Symbols indicate: □, $Re_b = 2205$; △, $Re_b = 3000$; $*$, $Re_b = 5000$. Black dashed line indicates a reference resolution $\Delta y/\eta = \pi/2$ commonly employed in isotropic turbulence studies; red dashed and dotted-dash lines indicate coarser $\Delta y/\eta = 2\sqrt{2}\pi/3$ and finer $\Delta y/\eta = \sqrt{2}\pi/3$ resolutions employed in Ishihara et al. (2007) respectively. Blue dashed line indicates the maximum ratio $\Delta y/\eta \approx 1.886$ for turbulent plane channel flow with $Re_\tau = 180$ of Hoyas and Jiménez (2008). Both data are of the duct bottom-wall bisector $z/H = 0$.

For completeness, the distribution of the local Kolmogorov length-scale η along the duct mid-span (cf. Figure A.2), as well as the grid resolutions in the three spatial directions directions in terms of the local η were also evaluated for three representative Reynolds numbers of $Re_b = 2205$, 3000 and 5000 (cf. Figure A.3).

Sufficiency of the streamwise extension $L_x/H = 8\pi$ was also examined in terms of two-point correlation of the normal components of Reynolds stresses for the open duct flow with $Re_b = 2205$ and $A = 1$ (cf. Figure A.4).

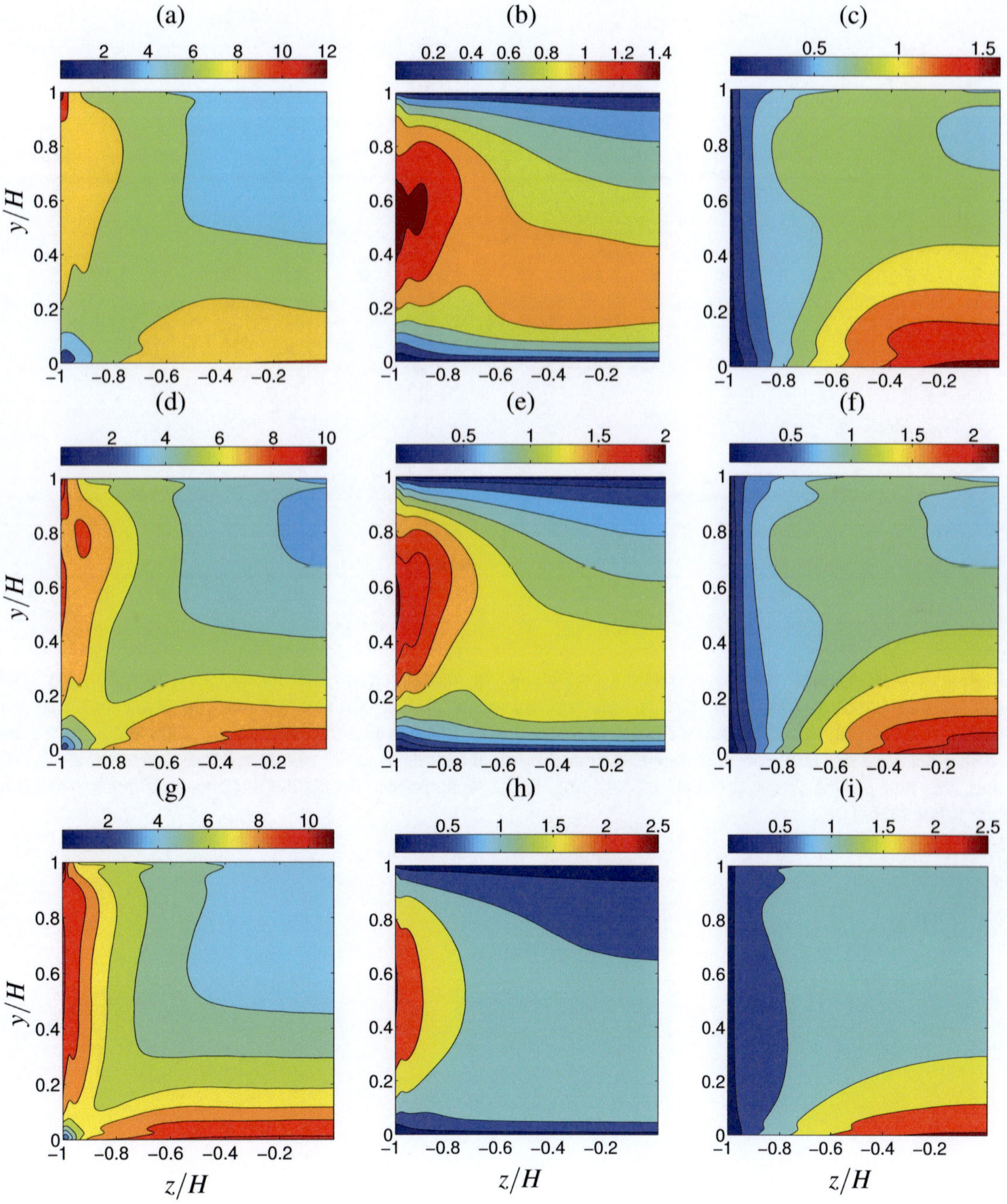

Figure A.3: Ratio of the grid spacing Δx_i to the local η, in: (a,d,g) streamwise; (b,e,h) wall-normal; (c,f,i) spanwise direction. (a-c) $Re_b = 2205$; (d-f) $Re_b = 3000$; (g-i) $Re_b = 5000$.

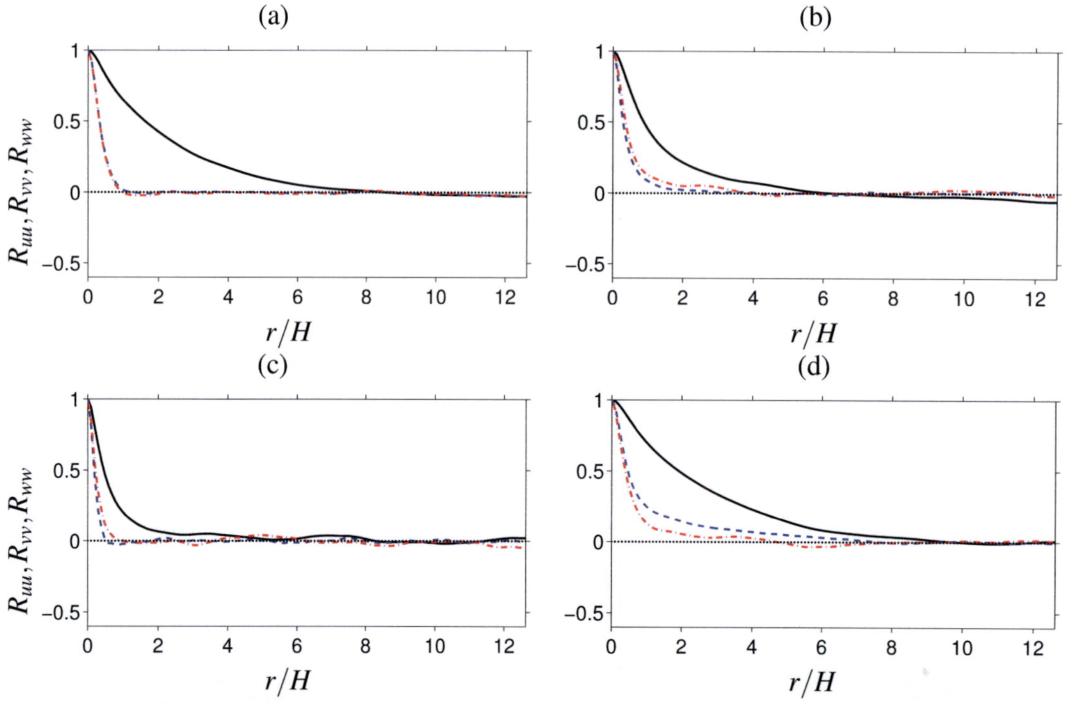

Figure A.4: Two-point velocity correlation coefficients of at four cross-plane locations $(y/H, z/H)$: (a) solid-solid corner $(0.059, -0.942)$; (b) bottom-wall duct mid-span $(0.059, -0.114)$; (c) free-slip plane duct mid-span $(0.876, -0.114)$; (d) mixed-boundary corner: $(0.876, -0.942)$, of open duct flow with $Re_b = 2205$ and $A = 1$. Those cross-plane locations (a-b) corresponding to the ones shown in Gavrilakis (1992). Different lines represent: black solid line, R_{uu}; blue dashed line, R_{vv}; red dotted dash line, Re_{ww}. Accumulated from 1000 instantaneous fields over a time interval of $2866H/u_b$.

Appendix B

Coherent structure analysis

Here we present the additional results of the two coherent structure analysis performed in Chapter 3, for the open duct flow with $A = 1$. First part of the analysis is the preferential locations of the quasi-streamwise vortex cores, educed by the technique based on the method proposed by Kida and Miura (1998). The flow bulk Reynolds numbers that are considered here are $Re_b = 1450$, 3000 and 5000, in Figure B.1, B.2 and B.3 respectively. For each case, the statistics were accumulated from 1000 snapshots separated by $10\nu/u_\tau^2$.

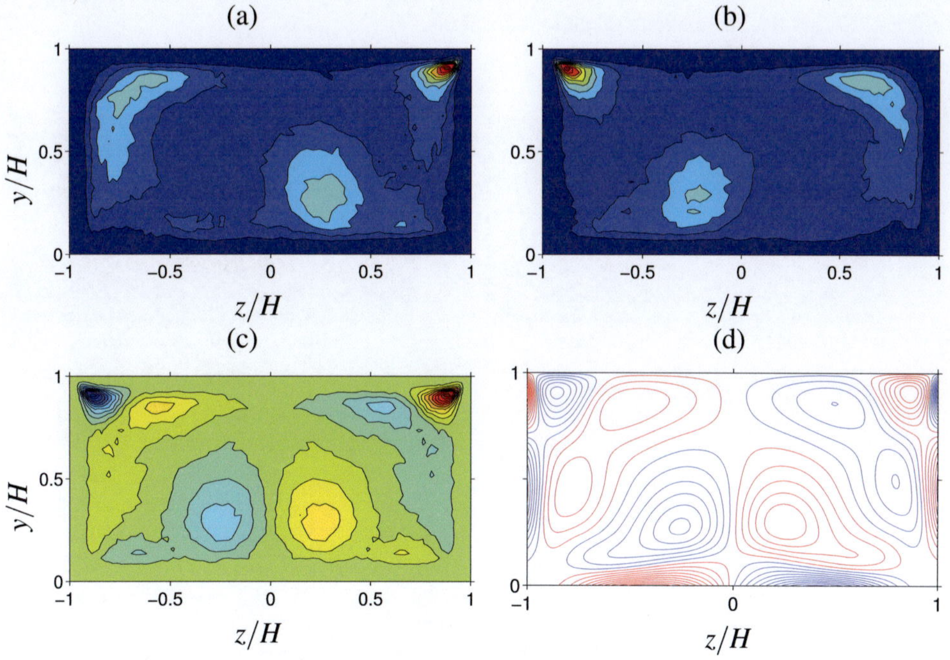

Figure B.1: (a-c) Probability of occurrence of vortex centres for the open duct case with $Re_b = 1450$, detected by the technique proposed by Kida and Miura (1998). (a) vortices with *positive* streamwise vorticity; (b) *negative* streamwise vorticity; (c) difference of (a) and (b). (c) half duct-width symmetrised. (d) Iso-contours of $\langle \omega_x \rangle (y,z)$ (replotted for convenience). The iso-countours indicate: (a,b) 0.1(0.1)0.9 times the maximum values; (c) -0.9(0.1)0.9 times the maximum values; (d) 0.1(0.05)0.9 times the maximum values respectively. Statistics were accumulated over 1000 snapshots over $\approx 10000\nu/u_\tau^2$, which is equivalent to $\approx 1366H/u_b$.

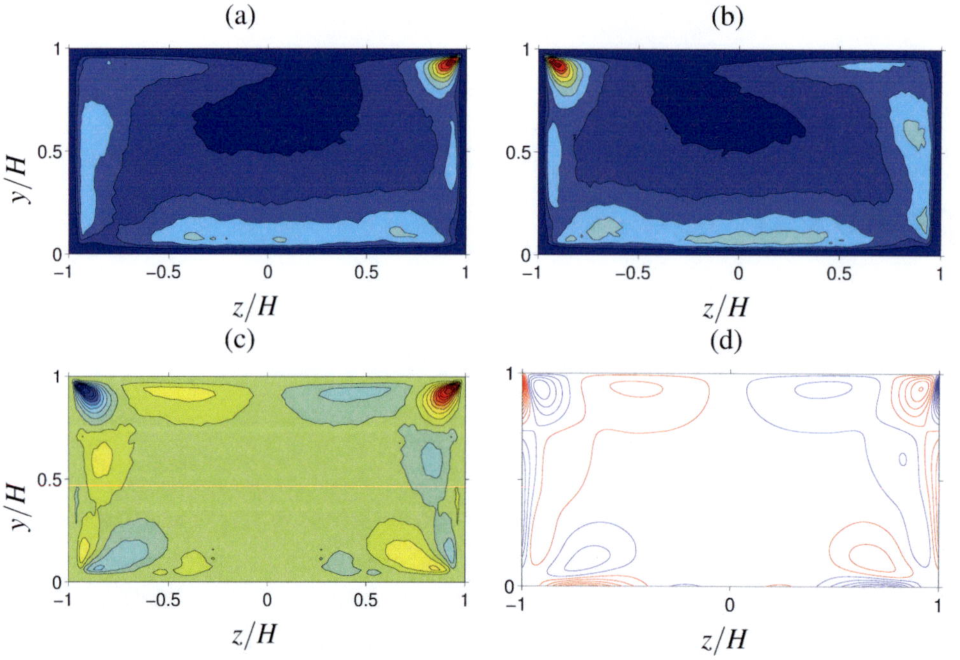

Figure B.2: (a-c) Probability of occurrence of vortex centres for the open duct case with $Re_b = 3000$, detected by the technique proposed by Kida and Miura (1998). (a) vortices with *positive* streamwise vorticity; (b) *negative* streamwise vorticity; (c) difference of (a) and (b). (c) half duct-width symmetrised. (d) Iso-contours of $\langle \omega_x \rangle (y,z)$ (replotted for convenience). The iso-countours indicate: (a,b) 0.1(0.1)0.9 times the maximum values; (c) -0.9(0.1)0.9 times the maximum values; (d) 0.1(0.05)0.9 times the maximum values respectively. Statistics were accumulated over 1000 snapshots over $\approx 10000\nu/u_\tau^2$, which is equivalent to $\approx 767H/u_b$.

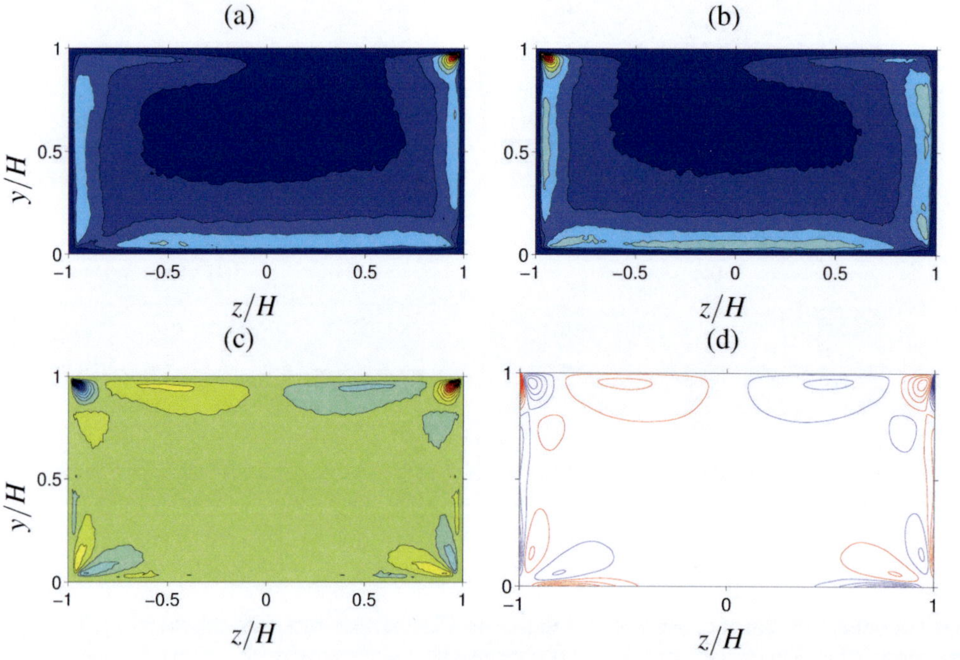

Figure B.3: (a-c) Probability of occurrence of vortex centres for the open duct case with $Re_b = 5000$, detected by the technique proposed by Kida and Miura (1998). (a) vortices with *positive* streamwise vorticity; (b) *negative* streamwise vorticity; (c) difference of (a) and (b). (c) half duct-width symmetrised. (d) Iso-contours of $\langle \omega_x \rangle (y,z)$ (replotted for convenience). The iso-countours indicate: (a,b) 0.1(0.1)0.9 times the maximum values; (c) -0.9(0.1)0.9 times the maximum values; (d) 0.1(0.05)0.9 times the maximum values respectively. Statistics were accumulated over 1000 snapshots over $\approx 10000\nu/u_\tau^2$, which is equivalent to $\approx 522H/u_b$.

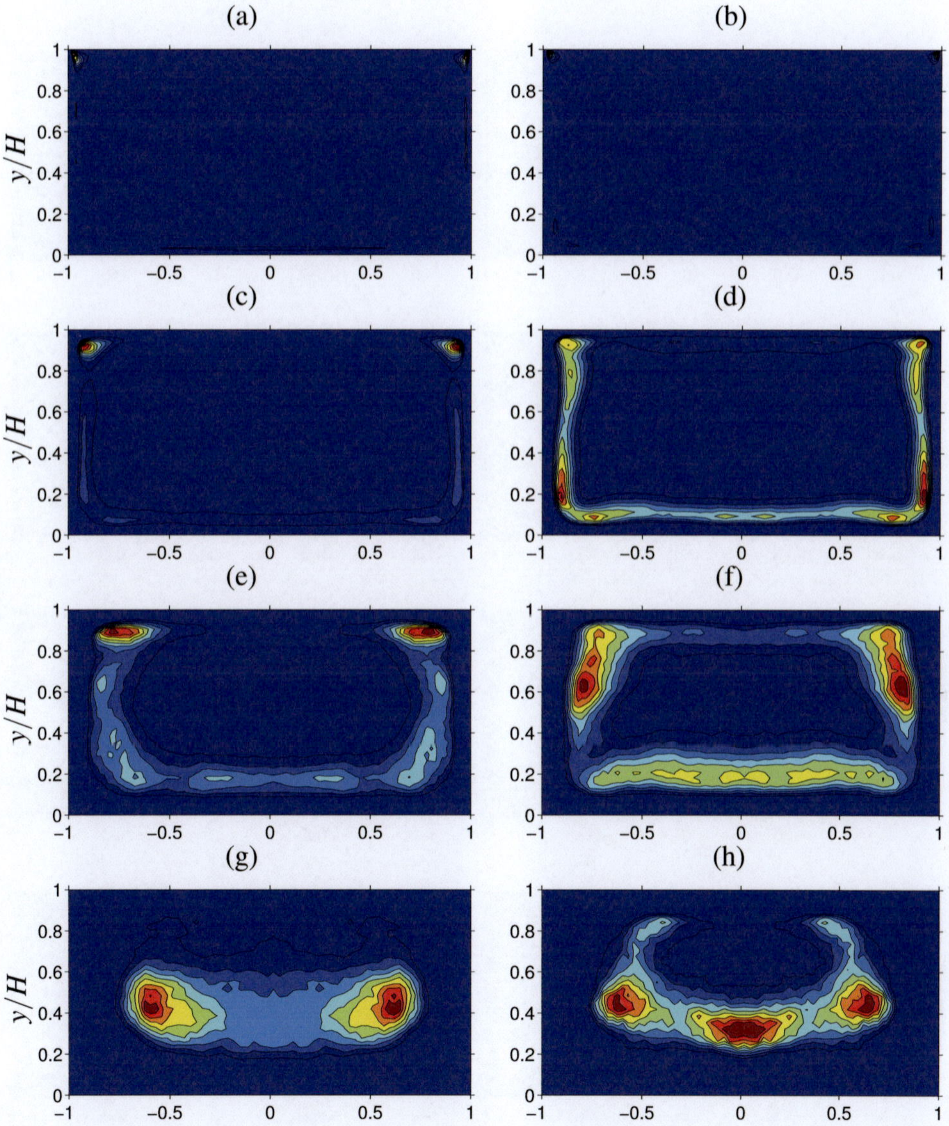

Figure B.4: Probability density function of the location of: (a,c,e,g) high-; (b,d,f,h) low-velocity streaks. Velocity streaks are classified by their length scale l_A as follows: (a,b) $l_A^+ < 25$; (c,d) $25 < l_A^+ < 50$; (e,f) $l_A^+ > 50$, $l_A/H < 0.5$; (g,h) $l_A/H > 0.5$. Contour lines indicate 0.1(0.1)0.9 times the maximum values, and statistics are symmetrised about the duct mid-span. Reynolds number is at $Re_b = 3000$, and the employed threshold value is at $u_{\text{th}} = 0.5$.

Second part of the analysis was the velocity streak eduction study based on the same snapshots. The flow bulk Reynolds numbers that are considered here are 3000 and 5000, shown respectively in Figure B.4 and B.5.

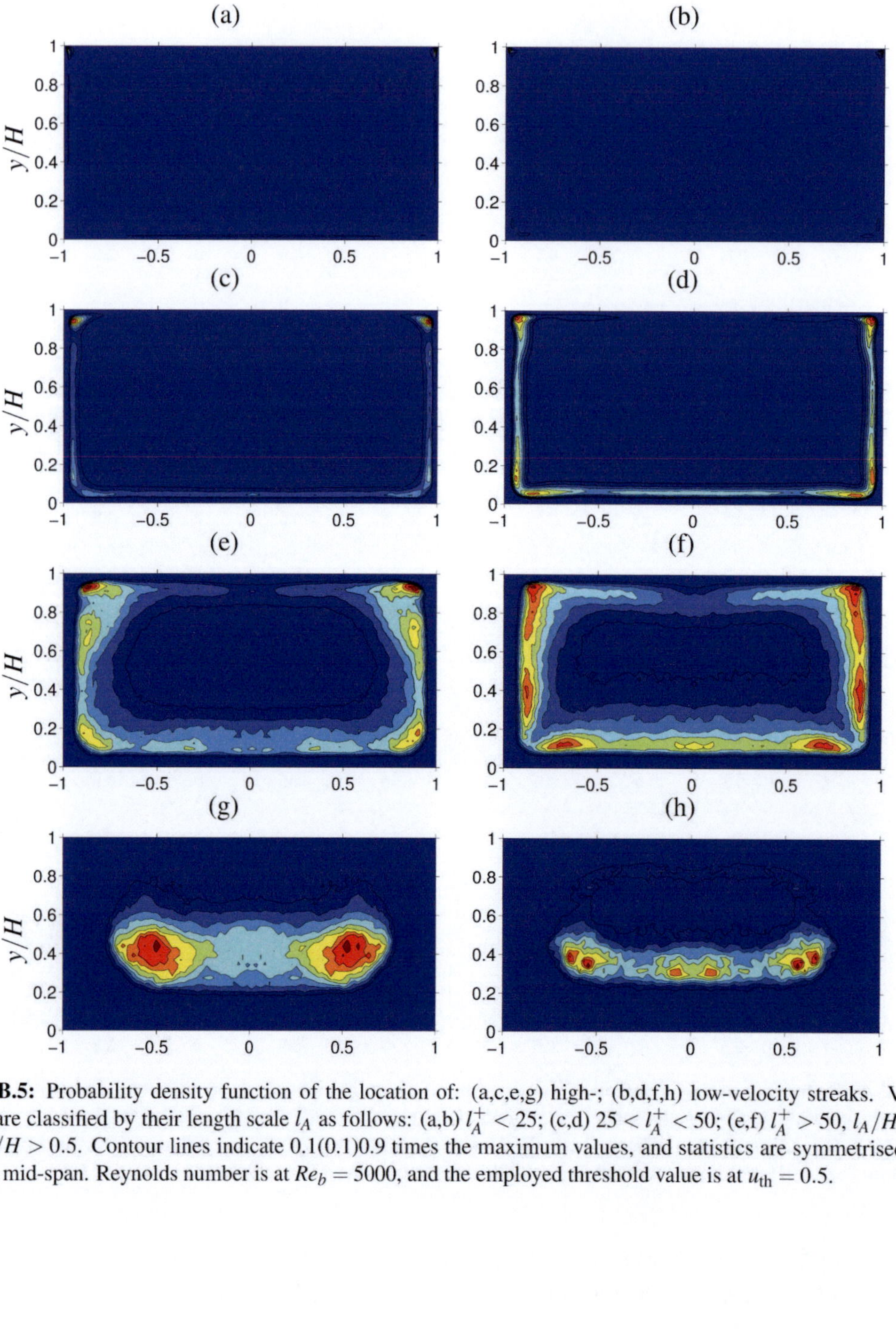

Figure B.5: Probability density function of the location of: (a,c,e,g) high-; (b,d,f,h) low-velocity streaks. Velocity streaks are classified by their length scale l_A as follows: (a,b) $l_A^+ < 25$; (c,d) $25 < l_A^+ < 50$; (e,f) $l_A^+ > 50$, $l_A/H < 0.5$; (g,h) $l_A/H > 0.5$. Contour lines indicate 0.1(0.1)0.9 times the maximum values, and statistics are symmetrised about the duct mid-span. Reynolds number is at $Re_b = 5000$, and the employed threshold value is at $u_{\text{th}} = 0.5$.

Appendix C

Improved fractional step method

C.1 Introduction

As it was mentioned in Chapter 2, the current pseudo-spectral method suffers from a severe spatial resolution requirement when the duct flow Reynolds number becomes relatively high, which makes infeasible to simulate the duct flow with Reynolds number higher than the ones considered within this thesis. At the time of writing, the numerical instability is suspected to be due to the spurious pressure modes, arisen from the collocated grid arrangement in the two-dimensional Chebyshev expansions and exist in each Fourier wavenumber, interacting with the no-slip walls and triggering the amplification of the erroneous slip-velocity on those boundaries.

The improved projection scheme of Hugues and Randriamampianina (1998) was therefore implemented, with a hope that it resolves the numerical instability. In the following, the improved scheme will be referred to as "**Hugues**", in comparison to our standard projection scheme denoted by "**OLD**" scheme.

The implementation was realised first in MATLAB, and later in FORTRAN (included in our pseudo-spectral DNS code), and the results from both implementations appeared to be consistent. Note that Hugues and Randriamampianina (1998) utilised a semi-implicit temporal-integration scheme consisting of Adams-Bashforth and backward differentiation formula (denoted by "**AB/BDF**"), whereas our original pseudo-spectral DNS code utilises low-storage three-step Runge-Kutta and Crank-Nicolson formulations. Both temporal schemes were implemented and validated accordingly.

Although the implementation was successfully validated against the test cases presented in Hugues and Randriamampianina (1998), the promised third-order accuracy in the slip-velocity error could not be achieved in the closed duct simulations, therefore the scheme was not employed in the production runs.

C.2 Validation

C.2.1 Analytical time-dependent Navier-Stokes solutions

First part of the validation was performed against analytical solutions of the two-dimensional Navier-Stokes equations induced by a forcing. First, we consider a steady solution which is followed by an unsteady problem. The analytical velocity and pressure fields are given as:

$$V(x,y,t) = \begin{pmatrix} \cos(\beta t)\sin(\frac{\pi x}{2})\cos(\frac{\pi y}{2}) \\ -\cos(\beta t)\cos(\frac{\pi x}{2})\sin(\frac{\pi y}{2}) \end{pmatrix} \tag{C.1}$$

$$p(x,y,t) = \cos^2(\beta t)[\cos(\pi x) + \cos(\pi y)]/4 + (x+y)\alpha\cos(\beta t) \tag{C.2}$$

which are induced by the following forcing terms:

$$F(x,y,t) = \begin{pmatrix} -\frac{1}{Re}\left[\beta Re\sin(\beta t) - \frac{\pi^2}{2}\cos(\beta t)\right]\sin(\frac{\pi x}{2})\cos(\frac{\pi y}{2}) + \alpha\cos(\beta t) \\ \frac{1}{Re}\left[\beta Re\sin(\beta t) - \frac{\pi^2}{2}\cos(\beta t)\right]\cos(\frac{\pi x}{2})\sin(\frac{\pi y}{2}) + \alpha\cos(\beta t) \end{pmatrix} \tag{C.3}$$

where α and β are pre-specified values. For $\alpha \neq 0$, the pressure gradient at boundaries has time-dependency. Two sets of configurations were considered, which differed by the values of α set at 0 and 10, while a fixed value of $\beta = 5$ as well as the number of Chebyshev modes $N = M = 33$ were used throughout. Those simulations were started from their analytical solutions ($t = 0$) and the simulated u, v and p were sampled during eight periods, $t \in [4\pi/5,\ 4\pi]$. Error of the computed solutions are measured in the following way:

$$E_{tot} = \max_t \left\{ \frac{1}{NM} \sum_{i=0}^{N} \sum_{j=0}^{M} [f(x_i,y_j,t) - f_{NM}(x_i,y_j,t)]^2 \right\}^{1/2} \tag{C.4}$$

where E_{tot} is the total error of the simulation domain, and f and f_{NM} are analytical as well as computed field values respectively. E_{tot} can be further decomposed into internal (E_{int}) and boundary (E_{bou}) contributions.

In all the test cases above, 0th-order temporal error convergence in pressure was observed both in Hugues-, as well as in OLD-scheme, which differ from what Hugues et al. reported. The error measure of pressure gradient, however, converges at the rate documented in the paper. This behaviour indicates that the computed pressure fields are experiencing so-called "spurious pressure modes".

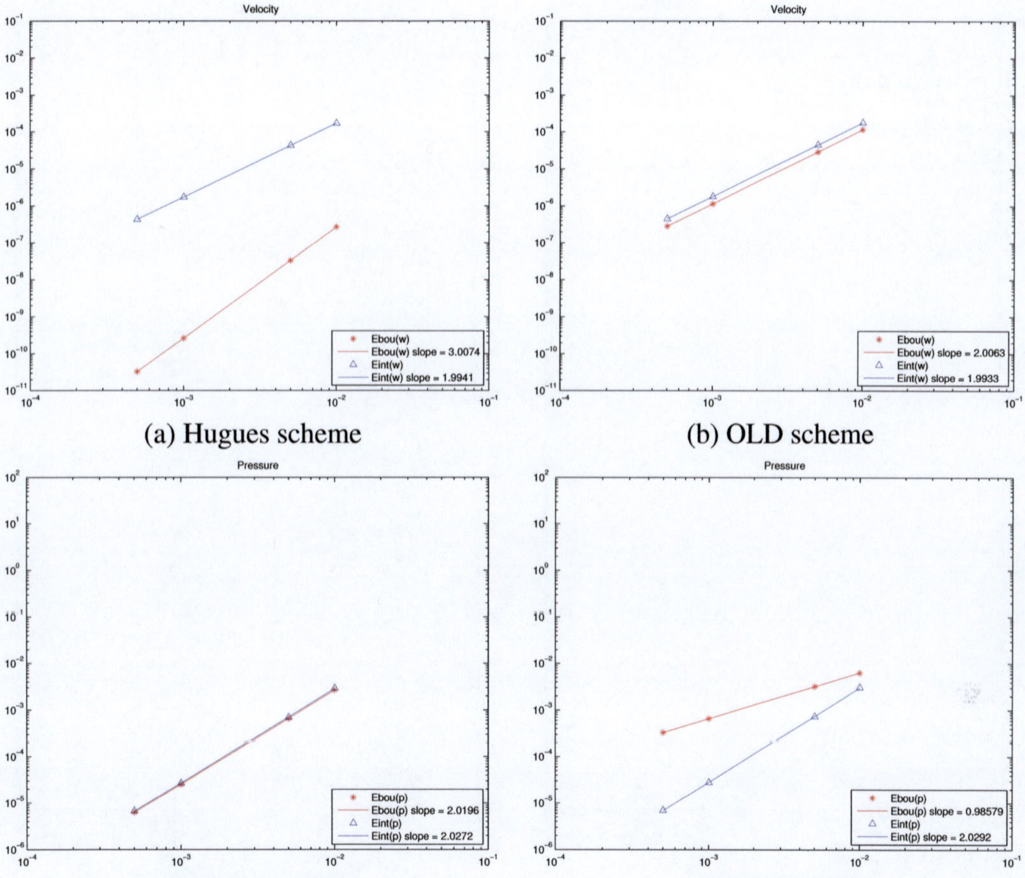

(a) Hugues scheme (b) OLD scheme

(c) Hugues scheme (d) OLD scheme

Figure C.1: Temporal convergence of L_2-error in: (a,b) velocity; and (c,d) pressure gradient. Analytical Navier-Stokes solutions with $\alpha = 0$, $\beta = 5$. (**MATLAB**)

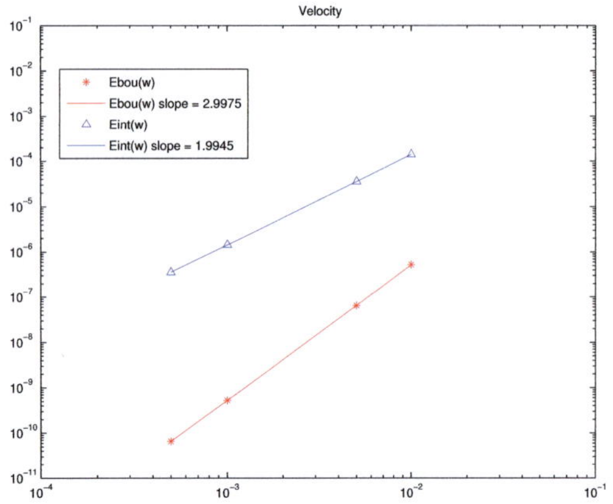

(a) Hugues scheme

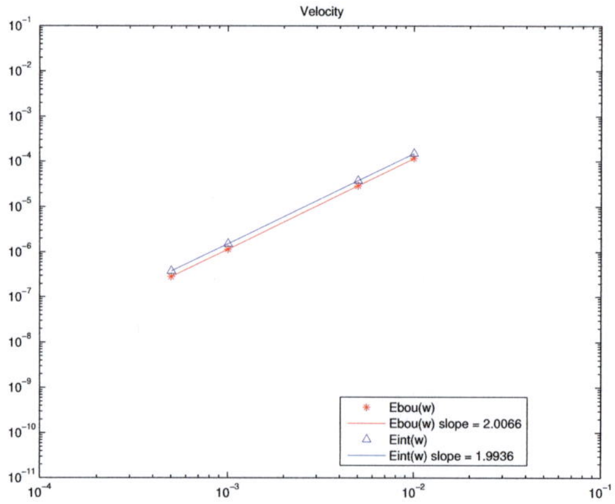

(b) OLD scheme

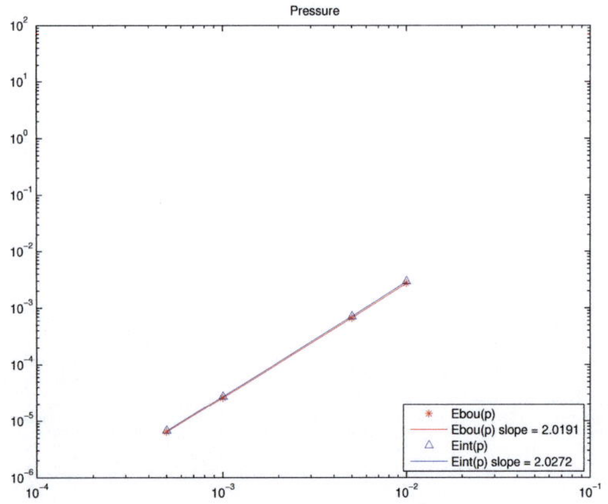

(c) Hugues scheme

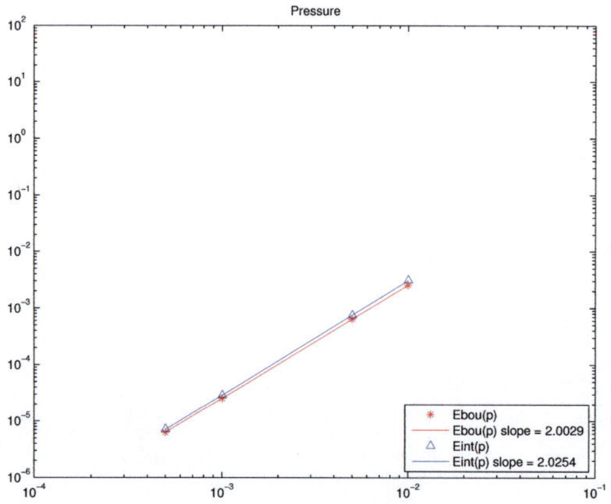

(d) OLD scheme

Figure C.2: Temporal convergence of L_2-error in: (a,b) velocity; and (c,d) pressure gradient. Analytical Navier-Stokes solutions with $\alpha = 0$, $\beta = 5$. (**FORTRAN**)

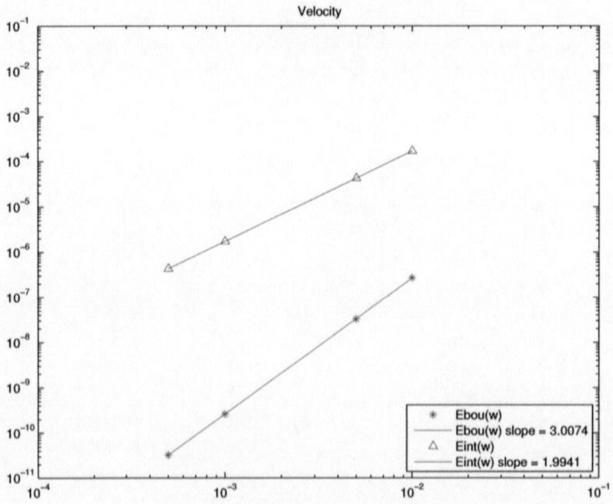

(a) Hugues scheme

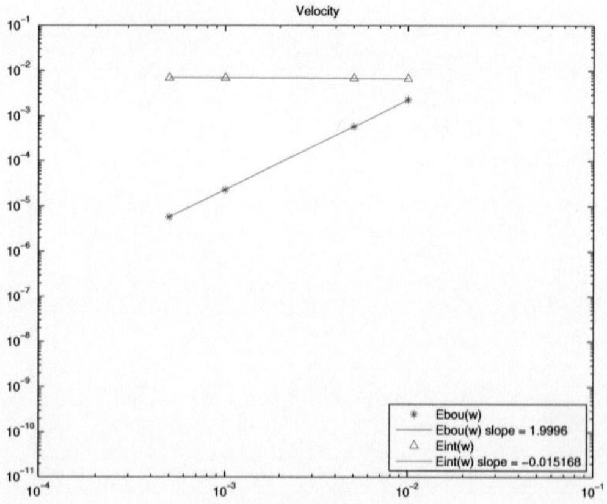

(b) OLD scheme

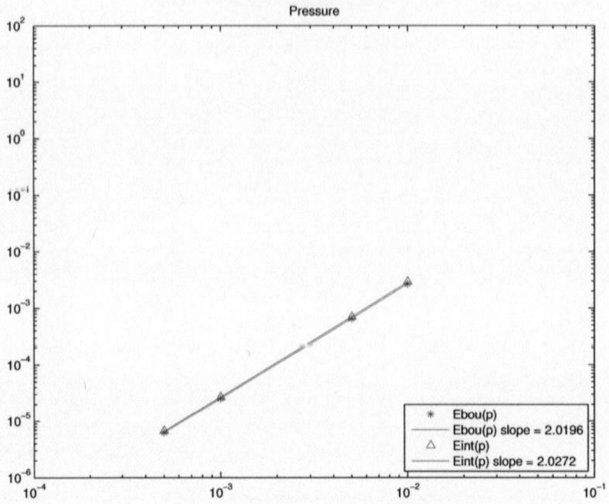

(c) Hugues scheme

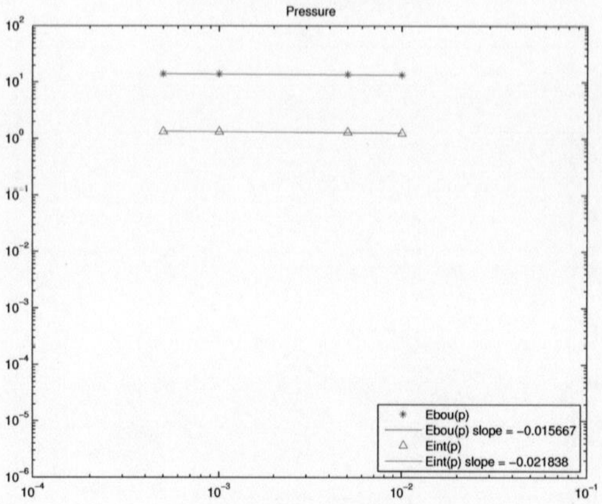

(d) OLD scheme

Figure C.3: Temporal convergence of L_2-error in: (a,b) velocity; and (c,d) pressure gradient. Analytical Navier-Stokes solutions with $\alpha = 10$, $\beta = 5$. (**MATLAB**)

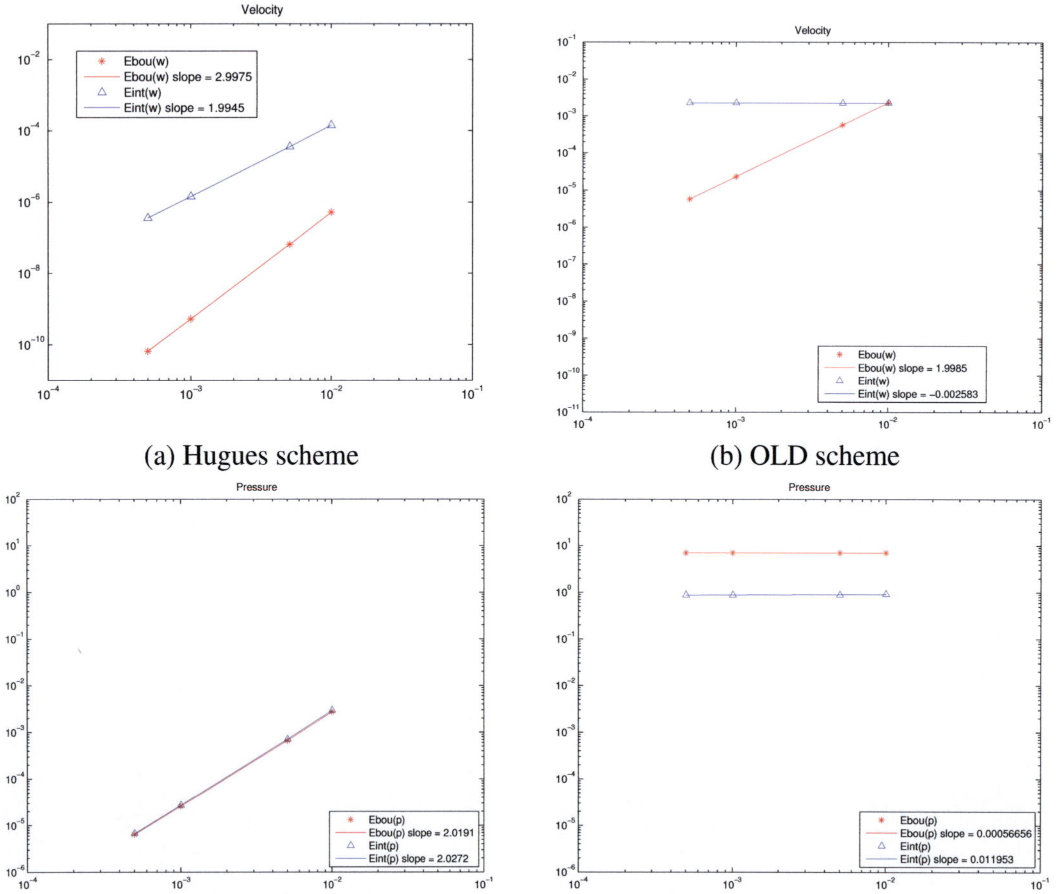

(a) Hugues scheme (b) OLD scheme

(c) Hugues scheme (d) OLD scheme

Figure C.4: Temporal convergence of L_2-error in: (a,b) velocity; and (c,d) pressure gradient. Analytical Navier-Stokes solutions with $\alpha = 10$, $\beta = 5$. (**FORTRAN**)

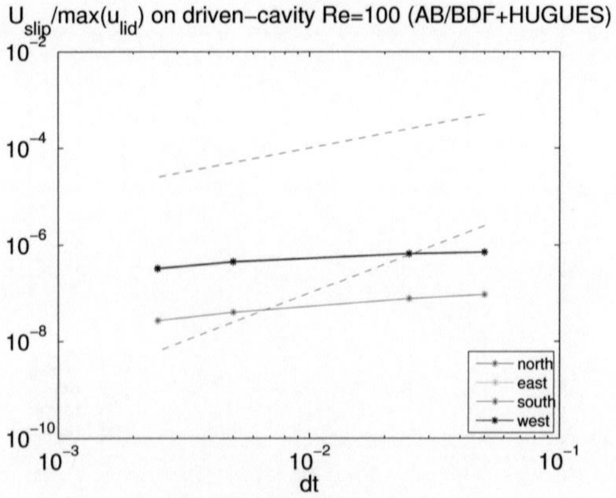

(a) AB/BDF+Hugues scheme

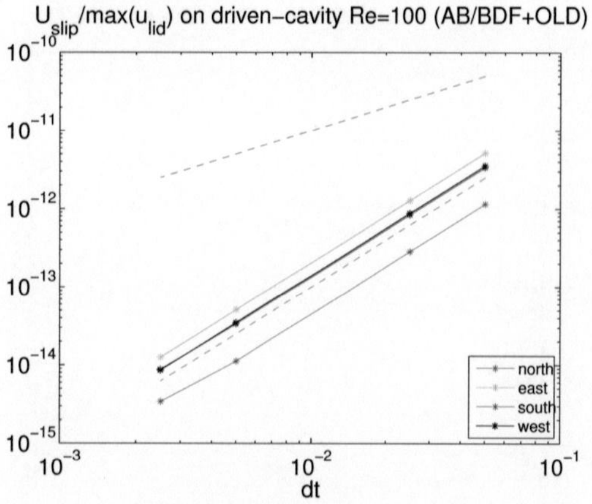

(b) AB/BDF+OLD scheme

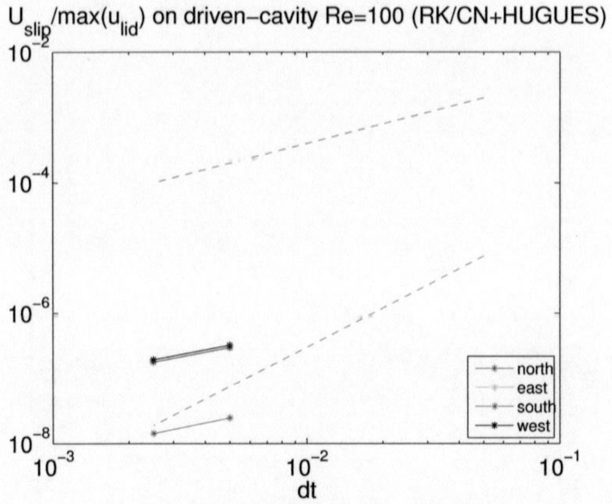

(a) RK/CN+Hugues scheme

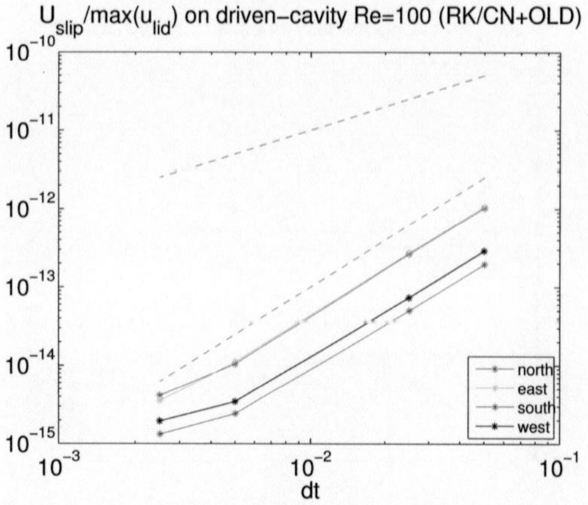

(b) RK/CN+OLD scheme

Figure C.5: Temporal convergence of L_2-error in slip-velocity normalised by the maximum lid velocity. "north", "east", "south" and "west" indicate individual walls, while red and blue bashed lines represent 1st and 2nd order temporal convergence. Used time-integration schemes are (a,b) AB/BDF; (c,d) RK/CN. (c) RK/CN has two less data points since the simulations diverged with those dt values. $M_x = 3$, $M_y = M_z = 33$. (**FORTRAN**)

C.2.2 Regularised lid-driven cavity

The second part of the validation was carried out against the two-dimensional regularised lid-driven driven cavity problem with Reynolds number at 100, which is again consistent with the validations shown in Hugues and Randriamampianina (1998). Temporal convergence of the slip-velocity error was evaluated, and presented in Figure C.5. Whilst a second-order convergence can be observed in the simulations by the OLD scheme, the Hugues scheme only shows 0-th order convergence.

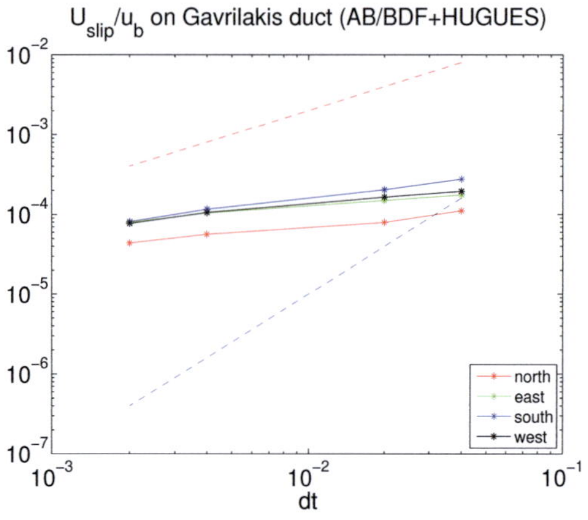

(a) AB/BDF+Hugues scheme

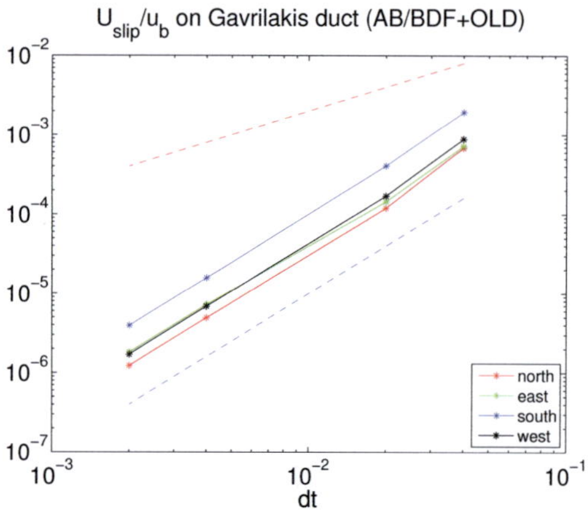

(b) AB/BDF+OLD scheme

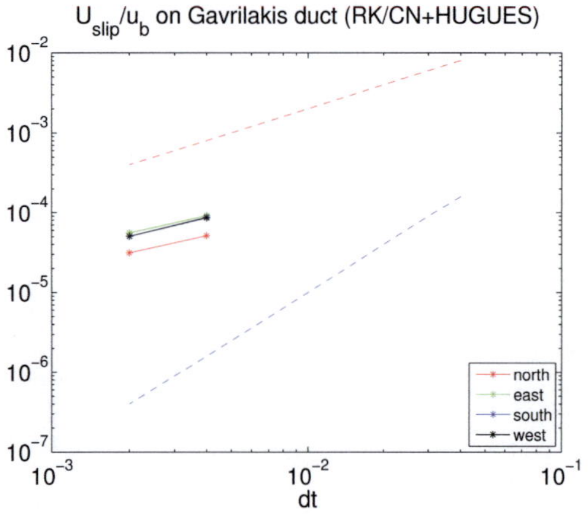

(c) RK/CN+Hugues scheme

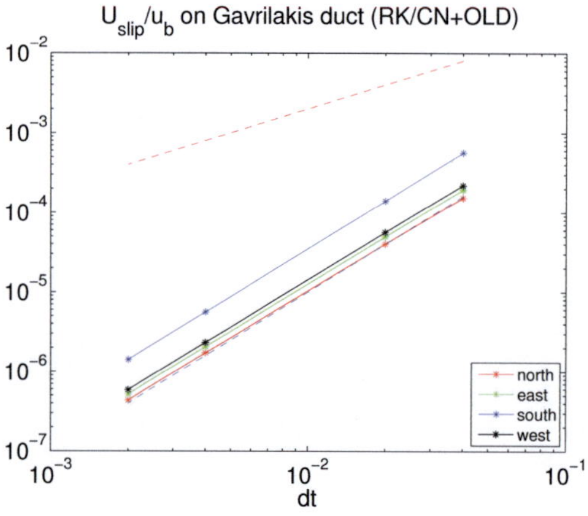

(d) RK/CN+OLD scheme

Figure C.6: Temporal convergence of L_2-error in slip-velocity normalised by bulk velocity. "north", "east", "south" and "west" indicate individual walls, while red and blue bashed lines represent 1st and 2nd order temporal convergence. Used time-integration schemes are: (a,b) AB/BDF (c,d) RK/CN. (c) RK/CN+HUGUES has two less data points since the simulations diverged with those *dt* values. $M_x = 256$, $M_y = M_z = 97$. (**FORTRAN**)

C.3 Closed duct turbulence (Gavrilakis)

Finally the Hugues scheme was validated against the square closed duct flow with $Re_b = 2205$. The employed measure is again the temporal convergence of the slip-velocity error, which is shown in Figure C.6. The results are almost identical to the above lid-driven cavity tests: Once again the Hugues scheme performed poorly with a 0th-order convergence, while the OLD scheme showed a consistent second-order convergence.

Appendix D

Inhomogeneous Neumann problem

More general form of Neumann boundary condition is *"inhomogeneous"* one. Similar to the homogeneous case, we use the a condensed matrix form that keeps into account the boundary conditions. Likewise Equations (2.21, 2.22), we start by considering the collocated inhomogeneous Neumann condition at the extrema of a horizontal line joining two vertical edges at location j. The left and right conditions are, respectively:

$$d_{0,0}\mathbf{U}_{0,j} + \sum_{i=1}^{N-1} d_{0,i}\mathbf{U}_{i,j} + d_{0,N}\mathbf{U}_{N,j} = a_j = \partial_x u|_{x=-1} \tag{D.1}$$

$$d_{N,0}\mathbf{U}_{0,j} + \sum_{i=1}^{N-1} d_{N,i}\mathbf{U}_{i,j} + d_{N,N}\mathbf{U}_{N,j} = b_j = \partial_x u|_{x=+1} \tag{D.2}$$

In a similar manner, we can write the Neumann condition at the extrema of a vertical line joining the horizontal edges of the square, as:

$$d_{0,0}\mathbf{U}_{i,0} + \sum_{j=1}^{N-1} d_{j,0}\mathbf{U}_{i,j} + d_{N,0}\mathbf{U}_{i,N} = c_i = \partial_y u|_{y=-1} \tag{D.3}$$

$$d_{0,N}\mathbf{U}_{i,0} + \sum_{j=1}^{N-1} d_{j,N}\mathbf{U}_{i,j} + d_{N,N}\mathbf{U}_{i,N} = e_i = \partial_y u|_{y=+1} \tag{D.4}$$

where $d_{i,j}$ are the entries of (2.12). Equations (D.1, D.2) can be solved in $\mathbf{U}_{0,j}, \mathbf{U}_{N,j}$, and Equations (D.3, D.4) can be solved in $\mathbf{U}_{i,0}, \mathbf{U}_{i,N}$. Therefore:

$$\mathbf{U}_{0,j} = \frac{d_{N,N}(a_j - \sum_{i=1}^{N-1} d_{0,i}\mathbf{U}_{i,j}) - d_{0,N}(b_j - \sum_{i=1}^{N-1} d_{N,i}\mathbf{U}_{i,j})}{d_{0,0}d_{N,N} - d_{N,0}d_{0,N}} \tag{D.5}$$

$$\mathbf{U}_{N,j} = \frac{-d_{N,0}(a_j - \sum_{i=1}^{N-1} d_{0,i}\mathbf{U}_{i,j}) + d_{0,0}(b_j - \sum_{i=1}^{N-1} d_{N,i}\mathbf{U}_{i,j})}{d_{0,0}d_{N,N} - d_{N,0}d_{0,N}} \tag{D.6}$$

$$\mathbf{U}_{i,0} = \frac{d^T_{N,N}(c_i - \sum_{j=1}^{N-1} d^T_{0,j}\mathbf{U}_{i,j}) - d^T_{0,N}(e_i - \sum_{j=1}^{N-1} d^T_{N,j}\mathbf{U}_{i,j})}{d_{0,0}d_{N,N} - d_{N,0}d_{0,N}} \tag{D.7}$$

$$\mathbf{U}_{i,N} = \frac{-d^T_{N,0}(c_i - \sum_{j=1}^{N-1} d^T_{0,j}\mathbf{U}_{i,j}) + d^T_{0,0}(e_i - \sum_{j=1}^{N-1} d^T_{N,j}\mathbf{U}_{i,j})}{d_{0,0}d_{N,N} - d_{N,0}d_{0,N}} \tag{D.8}$$

The obtained solution can be substituted in the (i,j) equation of system (2.13) to eliminate the boundary values. Operating in this way the i,j entry of the condensed form of matrix $\mathbf{D}^2$ again reads Equation 2.25 with the modified r.h.s. terms as:

$$\mathbf{F}^*_{i,j} = \mathbf{F}_{i,j} - \frac{\mathbf{T_1}_{i,j} + \mathbf{T_2}_{i,j} + \mathbf{T_3}_{i,j} + \mathbf{T_4}_{i,j}}{\mathbf{D}_{0,0}\,\mathbf{D}_{N,N} - \mathbf{D}_{0,N}\,\mathbf{D}_{N,0}} \tag{D.9}$$

where

$$\mathbf{T_1}_{i,j} = \mathbf{A}_j\,(\mathbf{D}^2_{i,0}\,\mathbf{D}_{N,N} - \mathbf{D}^2_{i,N}\,\mathbf{D}_{N,0}) \tag{D.10}$$

$$\mathbf{T_2}_{i,j} = \mathbf{B}_j\,(\mathbf{D}^2_{i,N}\,\mathbf{D}_{0,0} - \mathbf{D}^2_{i,0}\,\mathbf{D}_{1,N}) \tag{D.11}$$

$$\mathbf{T_3}_{i,j} = \mathbf{C}_i\,((\mathbf{D}^2_{0,j})^T\,\mathbf{D}_{N,N} - (\mathbf{D}^2_{N,j})^T\,\mathbf{D}_{N,0}) \tag{D.12}$$

$$\mathbf{T_4}_{i,j} = \mathbf{E}_i\,((\mathbf{D}^2_{N,j})^T\,\mathbf{D}_{0,0} - (\mathbf{D}^2_{0,j})^T\,\mathbf{D}_{1,N}) \tag{D.13}$$

for $i = 1..N-1$, $j = 1..N-1$. Once the condensed form of the $\mathbf{D}^2$ operator (2.25) as well as corrected internal data $\mathbf{F}^*$ have been established, the same procedure as for the Dirichlet case is carried out (i.e.: equations 2.19, 2.20 and 2.18) to obtain the values of the discrete solution of the internal points. To complete the solution phase we need to reconstruct the values of $\mathbf{U}$ on the boundary edges. To this end, we use equations (2.23, 2.24) for all the boundary nodes except for the four corner points where a special treatment needs to be used. Here we can just use four unknown values to fix a homogeneous Neumann condition. Same as in the homogeneous case, we define $\vec{n}$ as the $\pi/4$ direction oriented toward the inside of the square. This condition applied to the four corners lead to a 4×4 linear system of equations:

$$2\,\mathbf{D}_{0,0}\mathbf{U}_{0,0} + \mathbf{D}_{0,N}\mathbf{U}_{N,0} + \mathbf{D}_{0,N}\mathbf{U}_{0,N} = -\sum_{i=1,N-1}\mathbf{D}_{0,i}(\mathbf{U}_{i,0} + \mathbf{U}_{0,i}) + a_0 + c_0; \tag{D.14}$$

$$-\mathbf{D}_{N,0}\mathbf{U}_{0,0} + (\mathbf{D}_{0,0} - \mathbf{D}_{N,N})\mathbf{U}_{N,0} + \mathbf{D}_{0,N}\mathbf{U}_{N,N} = -\sum_{i=1,N-1}\mathbf{D}_{0,i}\mathbf{U}_{N,0} + \sum_{i=1,N-1}\mathbf{D}_{N,i}\mathbf{U}_{i,0} - b_0 + c_N; \tag{D.15}$$

$$2\,\mathbf{D}_{N,N}\mathbf{U}_{N,N} + \mathbf{D}_{N,0}\mathbf{U}_{0,N} + \mathbf{D}_{N,0}\mathbf{U}_{N,0} = -\sum_{i=1,N-1}\mathbf{D}_{N,i}(\mathbf{U}_{i,N} + \mathbf{U}_{N,i}) - B_N - e_N; \tag{D.16}$$

$$-\mathbf{D}_{N,0}\mathbf{U}_{0,0} + (\mathbf{D}_{0,0} - \mathbf{D}_{N,N})\mathbf{U}_{0,N} + \mathbf{D}_{0,N}\mathbf{U}_{N,N} = -\sum_{i=1,N-1}\mathbf{D}_{0,i}\mathbf{U}_{i,N} + \sum_{i=1,N-1}\mathbf{D}_{N,i}\mathbf{U}_{0,i} + a_N - e_0. \tag{D.17}$$

These equations are the discrete counterparts of the conditions:

$$\partial_x u + \partial_y u = a_0 + c_0 \quad \text{in } x = y = -1 \tag{D.18}$$

$$-\partial_x u + \partial_y u = -b_0 + c_N \quad \text{in } x = 1,\ y = -1 \tag{D.19}$$

$$-\partial_x u - \partial_y u = -b_N - e_N \quad \text{in } x = 1,\ y = 1 \tag{D.20}$$

$$\partial_x u - \partial_y u = a_N - e_0 \quad \text{in } x = -1,\ y = 1 \tag{D.21}$$

The linear system can be solved in the four corner values, once the value of the unknown on the four edges has been determined.
To eliminate the singularity of the Neumann problem in the case of with $c = 0$, the same techniques as the homogeneous case are employed.

D.1 Validation

Also for the Neumann case we made some numerical experiments using $c = 10$, and considering the exact solution:

$$u(x,y) = e^{(-cos(\pi x)^2)}\, e^{(-cos(\pi y))} \tag{D.22}$$

The following table summarizes the error convergence against the number of Chebyshev modes in homogeneous Neumann case.

nodes number	max. error
15×15	2.3×10^{-2}
30×30	3.2×10^{-5}
50×50	1.2×10^{-9}
90×90	7.5×10^{-12}

Similarly in inhomogeneous Neumann case:

nodes number	max. error	rms error
15×15	2.6×10^{-3}	9.4×10^{-4}
30×30	9.1×10^{-7}	4.2×10^{-7}
50×50	1.5×10^{-11}	6.6×10^{-12}
90×90	7.4×10^{-13}	2.8×10^{-13}

For $c = 0$ (Poisson problem):

nodes number	max. error	rms error
15×15	9.7×10^{-3}	3.2×10^{-3}
30×30	2.5×10^{-6}	1.7×10^{-6}
50×50	4.0×10^{-11}	2.7×10^{-11}
90×90	2.5×10^{-12}	1.6×10^{-12}

Please note that the original unit square domain was transformed into a rectangular shape by multiply x-coordinate by 0.9, y-coordinate by 1.1 in order to give inhomogeneous boundary conditions. To test against less-symmetric case, a new analytical test case was introduced by shifting x, y-coordinate by unity, i.e.:

$$u(x,y) = e^{(-cos(\pi x+1)^2)} \, e^{(-cos(\pi y+1))}. \tag{D.23}$$

Since our interest for the inhomogeneous solver is particularly on the Poisson problem, c is set at zero for the following test.

nodes number	max. error	rms error
15×15	4.4×10^{-2}	1.5×10^{-2}
30×30	8.7×10^{-5}	4.8×10^{-5}
50×50	1.6×10^{-9}	9.7×10^{-10}
90×90	2.6×10^{-12}	1.6×10^{-12}

Appendix E

Pseudo-spectral vorticity-streamfunction Navier-Stokes solver

E.1 Introduction

Our pseudo-spectral DNS technique employs a fractional time-stepping method to decouple the velocity and pressure computations in order to ease the required computing effort. The decoupling is realised by first computing the pseudo-velocity field and subsequently projecting it into the divergence-free space. Such projection procedure has a major drawback that is introducing so-called slip-velocity (or slip-error) around the domain boundaries. As a result, the method suffers from a lower-order accuracy in the boundary values of both velocity and pressure field.

Alternative approach to the above primitive variable formulation with the fractional time-stepping technique is to solve the governing Navier-Stokes equations in Vorticity-streamfunction ($\psi - \omega$) formulation. In this way, the solution is guaranteed to be free from the slip-error, whilst the continuity constraint is enforced by definition (cf. Ehrenstein and Peyret (1989)). There exits, however, the major challenge that is to define appropriate boundary conditions to the vorticity field.

Quartapelle and Valz-Gris (1981) proposed a semi-implicit scheme as a viable answer to the challenge, which couples the vorticity and streamfunction fields through their boundary conditions due to the implicit treatment of the diffusion terms. The coupled system requires, however, a rather complicated iterative method to be solved that is unfavourable from the associated computational effort viewpoint.

Consequently E and Liu (1996) approached to the high computational cost issue by modifying the formulation by Quartapelle and Valz-Gris (1981), by treating the diffusion terms explicitly. E and Liu (1996) demonstrated that the severe temporal resolution requirement arisen from the modification can be soften by employing third- or higher-order Runge-Kutta spatial discretization strategy.

E.2 Numerical procedure

Vorticity-stream function formulation of time-dependent 2D Navier-Stokes Equations can be written as:

$$\begin{aligned} \partial_t \omega + (\mathbf{u} \cdot \nabla)\omega &= \nu \Delta \omega, \\ \Delta \psi &= \omega \end{aligned} \tag{E.1}$$

with no-slip, impermeable boundary condition:

$$\psi|_\Gamma = 0, \; \left.\frac{\partial \psi}{\partial \mathbf{n}}\right|_\Gamma = 0$$

where $\mathbf{u} = (u, v)$; $u = -\partial_y \psi$, $v = \partial_x \psi$. Please note that definition of vorticity boundary condition is open as described in Section E.1.

E and Liu (1996) modified a method first introduced by Quartapelle and Valz-Gris (1981) with explicit treatment of the diffusion term, which discretizes the above equations as:

$$\begin{aligned} \frac{\omega^{n+1} - \omega^n}{\Delta t} + (\mathbf{u}^n \cdot \nabla)\omega^n &= \nu \Delta_h \omega^n \text{ for } i, j \geq 1, \\ \Delta_h \psi^{n+1} &= \omega^{n+1} \text{for } i, j \geq 1, \\ \psi|_\Gamma &= 0, \\ D_\mathbf{n} \psi^{n+1}|_\Gamma &= 0, \end{aligned} \tag{E.2}$$

where $D_\mathbf{n}$ is a gradient operator in the wall-normal directions, and $\Delta_h = D_x^2 D_y^2$ is a Laplace operator. For the computation, the scheme can be split into three steps:

Step 1: Update the vorticity at the interior grid points by:

$$\frac{\omega^{n+1} - \omega^n}{\Delta t} - \tilde{D}_y \psi^n \tilde{D}_x \omega^n + \tilde{D}_x \psi^n \tilde{D}_y \omega^n = \nu \Delta_h \omega^n, \tag{E.3}$$

where $\tilde{D}_*$ is a gradient operator for interior points in $*$ directions.

Step 2: Solve Poisson equation for ψ^{n+1} by:

$$\Delta \psi^{n+1} = \omega^{n+1}, \tag{E.4}$$

with the boundary condition $\psi|_\Gamma = 0$.

Step 3: Update the vorticity boundary values by computing the gradients of $\widehat{D_* \psi}$ whose boundary values are modified by velocity Dirichlet boundary conditions (cf. annotations written underneath of the below equations), thus:

$$\omega^{n+1}|_\Gamma = D_x(\underbrace{\widehat{D_x \psi^{n+1}}}_{v|_\Gamma}) + D_y(\underbrace{\widehat{D_y \psi^{n+1}}}_{-u|_\Gamma})|_\Gamma , \tag{E.5}$$

and

$$\omega^{n+1}|_{\text{corner}} = 0. \tag{E.6}$$

In case of Neumann boundary conditions (e.g. free-slip plane), the boundary values of $D_*(D_*\psi^{n+1})$ need to be specified instead of $D_*\psi^{n+1}$.
Explicit Euler time-integration scheme is used here only for demonstration purpose, and modifying to a higher-order Runge-Kutta scheme is straightforward. Before we move on to the next section, it is worth mentioning that the order of Runge-Kutta schemes reflects directly the number of Poisson problem necessary to be solved. In other words, 1st-order Euler solves one Poisson problems, while the 3rd and 4th order schemes solve 3 and 4 Poisson problems respectively.
Chebyshev polynomials are used for the spatial discretization, evaluated upon the Chebyshev-Gauss-Lobatto collocation points. Poisson problems are handled efficiently by a fast triagonalisation technique.

E.3 Validation

For validation, we considered two types of lid-driven cavities, namely *regularised cavity* and *singular cavity*, whose domains have an identical shape ($0 \leq x, y \leq 1$) but different lid velocity profiles. The lid velocity profile for the regularised cavity is defined as $u_{\text{lid}}(x) = 16x^2(1-x^2)$, whose velocity and its gradient smoothly decline to zero at the two lid corners. The singularity cavity has a uniform velocity profile defined as $u_{\text{lid}}(x) = 1$, which has singularity at the same corners due to the velocity discontinuity.
The presence of the singularity is particularly challenging to the codes based on spectral methods, whose global feature propagates any localised error throughout the simulation domain. Though several techniques were developed to overcome the singularity (e.g. subtraction method by Botella (1997)), no particular effort was made in the current implementation, therefore it is our interest to determine the size of the impact for future work.
For both cases, stream function field was initialised by:

$$\psi_0(x,y) = (y^2 - y^3)u_{\text{lid}}(x),$$

and the corresponding ω_0 field was computed by a Laplacian operation.
Reynolds number was calculated by:

$$Re = \frac{(u_{\text{lid}})_{\max} \cdot H}{\nu}$$

where $(u_{\text{lid}})_{\max}$ is the maximum value of the lid velocity profiles (unity for the both cases), H is the cavity height/width (also unity) and fluid kinematic viscosity ν. Please note that effective Re for the regularised cavity is lower compared to the singular cavity at the same Re, due to the lid velocity difference.

Rather strangely defined *discrete kinetic energy* was used for the analysis in various literatures (e.g. Shen (1991), Botella (1997), Botella and Peyret (2001), defined by

$$E(n\Delta t) = \left\{ \sum_{i,j=0}^{N-1} [(u_{i,j}^n)^2 + (v_{i,j}^n)^2] \right\}^{1/2}.$$

For the numerical stability from the (explicit) advection term of the algorithm, it is important to check CFL number, defined for 2D case as:

$$CFL = \left(\frac{u}{\Delta x} + \frac{v}{\Delta y} \right) \cdot \Delta t.$$

Also it is necessary to check β for the (explicit) diffusion term

$$\beta = \nu \left(\frac{\Delta t}{\Delta x^2} \right),$$

the value needed to be smaller than 0.5 from the linear stability analysis of advection-diffusion equations.

A criterion is necessary to determine the convergence of steady-state solutions and the normalised maximum residual of vorticity was used, which is defined as:

$$\frac{\max\limits_{i,j} |\omega_{i,j}^{n+1} - \omega_{i,j}^n|}{\Delta t \cdot \max\limits_{i,j} |\omega_{i,j}^{n+1}|} \leq \varepsilon$$

where threshold level ε varies in case-by-case basis.

In the following analysis, displayed figures are normalised by the cavity height/width H and the maximum lid velocity u_{lid}, unless stated specifically.

For time being, the code was written in MATLAB for proof-of-concept purpose and no particular attention was paid on optimisation of the code. There fore all the performance figures reported in this report are just for reference – the results will certainly vary if optimisation/implementation in faster programming language (e.g. FORTRAN) are done. All computations were curried out on our in-house TAIFUN cluster with single processor.

E.3.1 Regularised Lid-driven Cavity

E.3.1.1 Steady-state solution regime

We simulated regularised cavity at two moderate *Re* of 100 and 400: well-known steady-state regime and repeatedly analysed as benchmark (e.g. in Ehrenstein and Peyret (1989), Botella (1997)).

To determine the steady state, we set ε at 10^{-6} (as in Ehrenstein and Peyret, Botella used $\varepsilon = 2 \times 10^{-6}$), and number of Chebyshev modes N varied from 16 to 40, in order to compare with the results from the above-mentioned literatures.

Prior to the validation, a range of simulations were done in order to study: (1) effect of different time-integration schemes; (2) necessary size of Δt; and (3) associated numerical stability. For this

study, value of discrete kinetic energy E was used and we set 6-digits convergence as a sufficient level. The results are presented in Table E.1,E.2.
In summary:

- β seems to be the critical factor in stability for the simulated cases;
- 3rd-order scheme showed higher stability (please refer to Table E.3, E.4, E.5);
- $N \geq 24$ is sufficient for $Re = 100$, while $N \geq 32$ is necessary for $Re = 400$.

Three quantities were considered for validation, namely the maximum value of ψ on collocation points (M_1); the maximum value of ω on the upper side $y = 1$ (M_2); and the analogous value interpolated on 201 equidistant grid points (M_3). Chebyshev polynomial interpolation was performed for the purpose.

order	N	walltime/ts [s]	Δt	β	CFL	E
1	16	1.17×10^{-3}	0.017	2.85	0.08	0.360729483391582
			0.015	2.51	0.071	0.360729483613395
			0.01	1.68	0.047	0.360729484334376
			0.005	0.84	0.023	0.360729485445026
	20	1.65×10^{-3}	0.0065	2.80	0.04	0.360699619811284
			0.005	2.15	0.03	0.360699620246051
			0.0025	1.08	0.015	0.360699620484711
			0.00125	0.537	0.0075	0.360699620635796
	24	1.9×10^{-3}	0.0031	2.86	0.022	0.36069843679214
			0.0025	2.31	0.018	0.360698437032619
			0.00125	1.15	0.009	0.360698437159767
			0.000625	0.58	0.005	0.360698437236959
	32	2.8×10^{-3}	0.0009	2.73	0.009	0.360698330802022
			0.000625	1.90	0.006	0.360698330888029
			0.0003125	0.95	0.003	0.360698330926182
			0.00015625	0.48	0.002	0.360698330983546
3	16	2.4×10^{-3}	0.019	3.18	0.09	0.360732323104661
			0.0095	1.59	0.045	0.360730838715383
			0.0047	0.79	0.022	0.360730190881698
	20	3.5×10^{-3}	0.0074	3.18	0.04	0.360699948867609
			0.0037	1.59	0.022	0.360699773606774
	24	4.0×10^{-3}	0.00358	3.30	0.026	0.360698537879287
	32	6.7×10^{-3}	0.00109	3.31	0.01	0.360698341208350
4	16	3.0×10^{-3}	0.0175	2.93	0.082	0.360732536425197
			0.008	1.34	0.037	0.360730680867945
			0.004	0.67	0.019	0.360730080756470
	20	4.2×10^{-3}	0.0065	2.80	0.039	0.360699895890180
			0.0032	1.38	0.019	0.360699711655129
	24	5.0×10^{-3}	0.0031	2.86	0.023	0.360698480904179
	32	8.0×10^{-3}	0.0008	2.89	0.0095	0.360698331767589

Table E.1: Regularised lid-driven cavity at $Re = 100$. Δt not normalised.

order	N	walltime/ts [s]	Δt	β	CFL	E
1	16	1.17×10^{-3}	0.06	2.5	0.29	0.369712375726645
			0.03	1.26	0.14	0.369712352772915
	20	1.65×10^{-3}	0.025	2.69	0.15	0.366672966612219
			0.012	1.29	0.072	0.366672959765287
	24	1.9×10^{-3}	0.012	2.77	0.0875	0.3665975720594
			0.006	1.38	0.044	0.366597570455088
	32	2.8×10^{-3}	0.0037	2.81	0.036	0.366590825055718
			0.0018	1.37	0.018	0.366590824414347
	40	5.0×10^{-3}	0.0015	2.85	0.018	0.366590723911639
3	16	2.4×10^{-3}	0.075	3.14	0.36	0.369819265199165
			0.035	1.47	0.17	0.369781728129777
	20	3.5×10^{-3}	0.03	3.2	0.18	0.366677915861513
			0.015	1.6	0.9	0.366676042359027
	24	4.0×10^{-3}	0.014	3.2	0.1	0.366598210646047
			0.007	1.61	0.05	0.366597919292479
	32	6.7×10^{-3}	0.004	3.0	0.04	0.366590675189063
			0.002	1.5	0.02	0.366590676511083
	40	6.7×10^{-3}	0.0017	3.2	0.02	0.366590575737687
4	16	3.0×10^{-3}	0.07	2.93	0.34	0.369835022967294
			0.035	1.47	0.17	0.369788424925228
	20	4.2×10^{-3}	0.027	2.9	0.16	0.366678480115753
			0.0135	1.45	0.08	0.366676179049245
	24	5.0×10^{-3}	0.0126	2.9	0.09	0.366598415621058
	32	8.0×10^{-3}	0.0038	2.9	0.037	0.366590821176566
	40	8.0×10^{-3}	0.00152	2.89	0.019	0.366590722374529

Table E.2: Regularised lid-driven cavity at $Re = 400$. Δt not normalised.

		$Re = 100$			$Re = 400$	
N	$(\Delta t)_c$	$(\beta)_c$	$(\Delta t)_{6-\text{digits}}$	$(\Delta t)_c$	$(\beta)_c$	$(\Delta t)_{6-\text{digits}}$
16	0.017	2.85	–	0.06	2.5	–
20	0.0065	2.80	–	0.025	2.69	–
24	0.0031	2.86	–	0.012	2.77	–
32	0.0009	2.73	–	0.0037	2.81	–
40	–	–	–	0.0015	2.85	–

Table E.3: 1st-order scheme. $(\beta)_c \approx 2.85$.

		$Re = 100$			$Re = 400$	
N	$(\Delta t)_c$	$(\beta)_c$	$(\Delta t)_{6-\text{digits}}$	$(\Delta t)_c$	$(\beta)_c$	$(\Delta t)_{6-\text{digits}}$
16	0.019	3.18	–	0.075	3.14	0.035
20	0.0074	3.18	–	0.030	3.20	0.015
24	0.00358	3.30	–	0.014	3.20	–
32	0.00109	3.31	–	0.004	3.00	–
40	–	–	–	0.0017	3.20	–

Table E.4: 3rd-order scheme. $(\beta)_c \approx 3.2 - 3.3$.

	Re = 100			Re = 400		
N	$(\Delta t)_c$	$(\beta)_c$	$(\Delta t)_{6-\text{digits}}$	$(\Delta t)_c$	$(\beta)_c$	$(\Delta t)_{6-\text{digits}}$
16	0.075	2.93	0.008	0.070	2.93	0.035
20	0.0065	2.80	–	0.027	2.90	0.0135
24	0.0031	2.86	–	0.0126	2.90	–
32	0.0008	2.89	–	0.0038	2.90	–
40	–	–	–	0.00152	2.89	–

Table E.5: 4th-order scheme. $(\beta)_c \approx 2.9$.

N	M_1	M_1^E	M_1^{BS}	M_1^{BW}
16	8.1996×10^{-2}	8.3159×10^{-2}	8.3160×10^{-2}	8.3160×10^{-2}
	(0.35-0.75)	(0.40-0.78)	(0.40-0.78)	(0.40-0.78)
20	8.3251×10^{-2}	8.2695×10^{-2}	8.2694×10^{-2}	8.2695×10^{-2}
	(0.38-0.77)	(0.42-0.73)	(0.42-0.73)	(0.42-0.73)
24	8.3270×10^{-2}	8.3315×10^{-2}	8.3315×10^{-2}	8.3315×10^{-2}
	(0.40-0.73)	(0.37-0.75)	(0.37-0.75)	(0.37-0.75)
32	8.3393×10^{-2}	8.3402×10^{-2}	8.3402×10^{-2}	8.3402×10^{-2}
	(0.37-0.76)	(0.40-0.74)	(0.40-0.74)	(0.40-0.74)

Table E.6: M_1 at $Re = 100$

N	M_2	M_2^E	M_2^{BS}	M_2^{BW}
16	13.1815 (0.65)	13.3687 (0.60)	13.3467 (0.60)	13.3475 (0.60)
20	13.4427 (0.62)	13.1780 (0.66)	13.1759 (0.65)	13.1758 (0.65)
24	13.3778 (0.60)	13.4227 (0.63)	13.4226 (0.63)	13.4226 (0.63)
32	13.4369 (0.63)	13.3422 (0.60)	13.3423 (0.60)	13.3423 (0.60)

Table E.7: M_2 at $Re = 100$

N	M_3	M_3^E	M_3^{BS}	M_3^{BW}
16	13.4506 (0.620)	13.4663 (0.620)	13.4476 (0.620)	13.4464 (0.620)
20	13.4451 (0.620)	13.4459 (0.620)	13.4441 (0.620)	13.4433 (0.620)
24	13.4447 (0.620)	13.4446 (0.620)	13.4446 (0.620)	13.4445 (0.620)
32	13.4446 (0.620)	13.4447 (0.620)	13.4448 (0.620)	13.4447 (0.620)
40	–	–	13.4448 (0.620)	13.4447 (0.620)

Table E.8: M_3 at $Re = 100$

N	M_1	M_1^E	M_1^{BS}	M_1^{BW}
16	8.5447×10^{-2}	8.5378×10^{-2}	8.5777×10^{-2}	8.5470×10^{-2}
	(0.45-0.65)	(0.40-0.60)	(0.40-0.60)	(0.40-0.60)
20	8.5227×10^{-2}	8.5213×10^{-2}	8.5192×10^{-2}	8.5214×10^{-2}
	(0.46-0.62)	(0.43-0.58)	(0.42-0.58)	(0.42-0.58)
24	8.5462×10^{-2}	8.5715×10^{-2}	8.5716×10^{-2}	8.5715×10^{-2}
	(0.40-0.60)	(0.43-0.63)	(0.43-0.63)	(0.43-0.63)
32	8.5853×10^{-2}	8.5480×10^{-2}	8.5480×10^{-2}	8.5480×10^{-2}
	(0.42-0.63)	(0.40-0.60)	(0.40-0.60)	(0.40-0.60)

Table E.9: M_1 at $Re = 400$

N	M_2	M_2^E	M_2^{BS}	M_2^{BW}
16	25.0608 (0.65)	25.2329 (0.60)	24.7799 (0.60)	24.6615 (0.60)
20	24.9443 (0.62)	24.6693 (0.65)	24.6268 (0.65)	24.6176 (0.65)
24	24.6302 (0.60)	24.9344 (0.63)	24.9157 (0.63)	24.9155 (0.63)
32	24.9110 (0.63)	24.7845 (0.65)	24.7845 (0.65)	24.7845 (0.65)

Table E.10: M_2 at $Re = 400$

N	M_3	M_3^E	M_3^{BS}	M_3^{BW}
16	25.4354 (0.625)	25.4675 (0.620)	25.1604 (0.625)	25.0302 (0.625)
20	24.9515 (0.625)	24.9846 (0.630)	24.9273 (0.630)	24.9182 (0.625)
24	24.9125 (0.625)	24.9333 (0.630)	24.9148 (0.630)	24.9146 (0.630)
32	24.9110 (0.630)	24.9110 (0.630)	24.9111 (0.630)	24.9110 (0.630)
40	24.9109 (0.630)	–	24.9109 (0.630)	24.9109 (0.630)

Table E.11: M_3 at $Re = 400$

order	$(\Delta t)_c$	CFL	β	walltime[%]	$T_{\text{converged}}$	E_{final}
1	0.0025	0.059	1.2943	100	3412.86	0.375626267225189
3	0.005	0.12	2.5886	264	1925.99	0.375626144177793
4	0.005	0.12	2.5886	334	1811.81	0.375626134916356

Table E.12: Preliminary tests on $Re = 10000$. $(\Delta t)_c$ is the largest Δt reached to the steady-state (not normalised), and E_{final} is E at that point.

Case	N	max(E)	min(E)
Present	64	0.37519	0.37496
Shen	48	0.37548	0.37525
Shen	64	0.37521	0.37499

Table E.13: $Re = 10500$

E.3.1.2 Periodic solution regime

Shen (1991) showed that the regularised cavity solution becomes periodic by Hopf bifurcation somewhere in the interval of]10000,10500], and computed $Re = 10500$ cavity for demonstration. Later, Botella (1997) showed that the bifurcation occurs as early as $Re = 10300$.

Here we computed $Re = 10500$ cavity by avoiding uncertainty from the Re very close to the bifurcation point, restarting from the steady-state solution obtained for $Re = 10000$ (with $\varepsilon = 2 \cdot 10^{-6}$ as in Shen (1991)). Parametric tests with the 3 different time-integration schemes (1st-order Euler, 3rd/4th order Runge-Kutta) were performed for the $Re = 10000$ cavity to select a time order and the corresponding Δt for the periodic solution. The results are shown in Table E.12 and 3rd-order Runge-Kutta with $\Delta t = 0.0025$ was chosen.

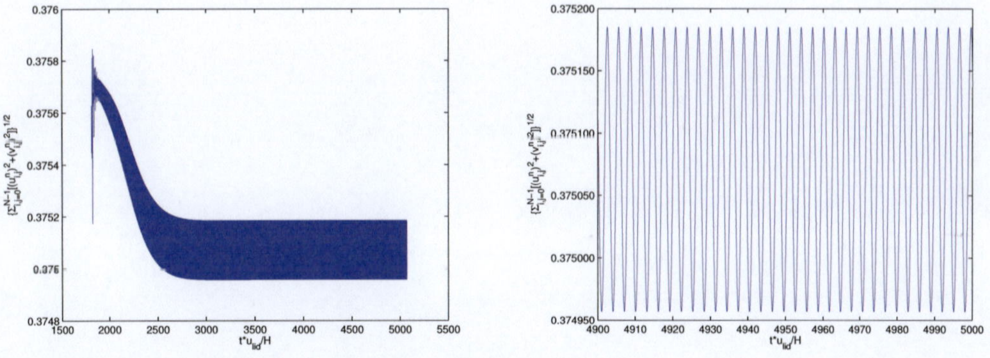

Figure E.1: Evolution of E for the regularised cavity at $Re = 10500$ starting from the steady-state solution of $Re = 10000$. Computed by 4th-order scheme. Oscillation period $T \approx 3$

Bibliography

Javier Jiménez. How linear is wall-bounded turbulence? *Phys. Fluids*, 25(11):110814, 2013. doi:10.1063/1.4819081.

S Gavrilakis. Numerical simulation of low-Reynolds-number turbulent flow through a straight square duct. *J. Fluid Mech.*, 244(-1):101, apr 1992. doi:10.1017/S0022112092002982.

Markus Uhlmann, Alfredo Pinelli, Genta Kawahara, and Atsushi Sekimoto. Marginally turbulent flow in a square duct. *J. Fluid Mech.*, 588(2000):153–162, sep 2007. doi:10.1017/S0022112007007604.

Alfredo Pinelli, Markus Uhlmann, Atsushi Sekimoto, and Genta Kawahara. Reynolds number dependence of mean flow structure in square duct turbulence. *J. Fluid Mech.*, 644:107, feb 2010. doi:10.1017/S0022112009992242.

Osborne Reynolds. An Experimental Investigation of the Circumstances Which Determine Whether the Motion of Water Shall Be Direct or Sinuous, and of the Law of Resistance in Parallel Channels. *Philos. Trans. R. Soc. London*, 174(1883):935–982, 1883. doi:10.1098/rstl.1883.0029.

Osborne Reynolds. On the dynamical theory of incompressible viscous fluids and the determination of the criterion. *Philos. Trans. R. Soc. London*, 186(1895):123–164, 1895. doi:10.1098/rsta.1979.0079.

Milton Van Dyke. *An Album of Fluid Motion*. THE PARABOLIC PRESS, 1982. ISBN 978-0-915760-03-9.

J. C. R. Hunt, A. A. Wray, and Parviz Moin. Eddies, Streams, and Converence Zones in Turbulent Flows. In *Cent. Turbul. Res.*, number Report CTR-S88 (1988), 1988.

Keisuke Takeishi, Genta Kawahara, Hiroki Wakabayashi, Markus Uhlmann, and Alfredo Pinelli. Localized turbulence structures in transitional rectangular-duct flow. *J. Fluid Mech.*, 782:368–379, 2015. doi:10.1017/jfm.2015.546.

Björn Hof, Jerry Westerweel, Tobias M Schneider, and Bruno Eckhardt. Finite lifetime of turbulence in shear flows. *Nature*, 443(7107):59–62, 2006. doi:10.1038/nature05089.

D Barkley, B Song, V Mukund, G Lemoult, M. Avila, and Björn Hof. The rise of fully turbulent flow. *Nature*, 526(7574):550–553, 2015. doi:10.1038/nature15701.

Masaki Sano and Keiichi Tamai. A universal transition to turbulence in channel flow. *Nat. Phys.*, 12(3):249–253, feb 2016. doi:10.1038/nphys3659.

Peter J. Schmid and Dan S. Henningson. *Stability and Transition in Shear Flows*, volume 142 of *Applied Mathematical Sciences*. Springer New York, New York, NY, 2001. ISBN 978-1-4612-6564-1. doi:10.1007/978-1-4613-0185-1.

Stephen B. Pope. *Turbulent Flows*. Cambridge University Press, 2000. ISBN 0521598869.

A. O. Demuren and Wolfgang Rodi. Calculation of turbulence-driven secondary motion in non-circular ducts. *J. Fluid Mech.*, 140(-1):189–222, 1984. doi:10.1017/S0022112084000574.

Lewis Fry Richardson. *Weather prediction by numerical process*. Cambridge University Press, 1922.

Andrey Nikolaevich Kolmogorov. The local structure of turbulence in incompressible viscous fluid for very large Reynolds numbers. *Dokl. Akad. Nauk SSSR*, 30:299–303, 1941.

Ludwig Prandtl. Über Flüssigkeitsbewegung bei sehr kleiner Reibung ("Motion of Fluids with Very Little Viscosity"). *Int. Math.*, pages 1–8, 1904.

John D. Anderson. Ludwig Prandtl ' s Boundary Layer. *Phys. Today*, (December):7, 2005. doi:10.1063/1.2169443.

Ludwig Prandtl. Bericht über die Entstehung der Turbulenz. *Zeitschrift für Angew. Math. und Mech.*, 5:136–139, 1925.

T. Wei and W. W. Willmarth. Reynolds-number effects on the structure of a turbulent channel flow. *J. Fluid Mech.*, 204(-1):57–95, apr 1989. doi:10.1017/S0022112089001667.

John Kim, Parviz Moin, and Robert D. Moser. Turbulence statistics in fully developed channel flow at low Reynolds number. *J. Fluid Mech.*, 177(-1):133–166, apr 1987. doi:10.1017/S0022112087000892.

Javier Jiménez and Genta Kawahara. Dynamics of Wall-Bounded Turbulence. In Peter A. Davidson, Yukio Kaneda, and Katepalli R. Sreenivasan, editors, *Ten Chapters Turbul.*, pages 221–268. Cambridge University Press, Cambridge, 2013. ISBN 9781139032810 (ISBN); 9780521769440 (ISBN). doi:10.1017/CBO9781139032810.007.

Alexander J. Smits, Beverley J. McKeon, and Ivan Marusic. High-Reynolds Number Wall Turbulence. *Annu. Rev. Fluid Mech.*, 43(1):353–375, jan 2011. doi:10.1146/annurev-fluid-122109-160753.

Michio Nishioka and Masahito Asai. Some observations of the subcritical transition in plane Poiseuille flow. *J. Fluid Mech.*, 150(-1):441–450, 1985. doi:10.1017/S0022112085000210.

Ricardo Vinuesa, Philipp Schlatter, and Hassan M. Nagib. On minimum aspect ratio for duct flow facilities and the role of side walls in generating secondary flows. *J. Turbul.*, 16(6):588–606, 2015. doi:10.1080/14685248.2014.996716.

S. J. Kline, W. C. Reynolds, F. A. Schraub, and P. W. Runstadler. The structure of turbulent boundary layers. *J. Fluid Mech.*, 30(04):741–773, 1967. doi:10.1017/S0022112067001740.

H. T. Kim, S. J. Kline, and W. C. Reynolds. The production of turbulence near a smooth wall in a turbulent boundary layer. *J. Fluid Mech.*, 50(01):133–160, 1971. doi:10.1017/S0022112071002490.

Philippe R Spalart. Direct simulation of a turbulent boundary layer up to Rθ = 1410. *J. Fluid Mech.*, 187:61–98, 1988. doi:10.1017/S0022112088000345.

C. R. Smith and S. P. Metzler. The characteristics of low-speed streaks in the near-wall region of a turbulent boundary layer. *J. Fluid Mech.*, 129(-1):27, apr 1983. doi:10.1017/S0022112083000634.

Jinhee Jeong, Fazle Hussain, W. Schoppa, and John Kim. Coherent structures near the wall in a turbulent channel flow. *J. Fluid Mech.*, 332:185–214, 1997.

Javier Jiménez and Parviz Moin. The minimal flow unit in near-wall turbulence. *J. Fluid Mech.*, 225, 1991.

MJ Lee, John Kim, and Parviz Moin. Structure of turbulence at high shear rate. *J. Fluid Mech.*, 1990.

Ron F. Blackwelder and Helmut Eckelmann. Streamwise vortices associated with the bursting phenomenon. *J. Fluid Mech.*, 94(03):577, 1979. doi:10.1017/S0022112079001191.

G. R. Offen and S. J. Kline. A proposed model of the bursting process in turbulent boundary layers. *J. Fluid Mech.*, 70(02):209, 1975. doi:10.1017/S002211207500198X.

S K Robinson. Coherent Motions in the Turbulent Boundary Layer. *Annu. Rev. Fluid Mech.*, 23 (1):601–639, jan 1991. doi:10.1146/annurev.fl.23.010191.003125.

Javier Jiménez, Genta Kawahara, Mark P. Simens, Masato Nagata, and Makoto Shiba. Characterization of near-wall turbulence in terms of equilibrium and "bursting" solutions. *Phys. Fluids*, 17(1), 2005. doi:10.1063/1.1825451.

Jerry D. Swearingen and Ron F. Blackwelder. The growth and breakdown of streamwise vortices in the presence of a wall. *J. Fluid Mech.*, 182(-1):255, 1987. doi:10.1017/S0022112087002337.

Javier Jiménez and Alfredo Pinelli. The autonomous cycle of near-wall turbulence. *J. Fluid Mech.*, 389:335–359, 1999. doi:10.1017/S0022112099005066.

John Kim and Fazle Hussain. Propagation velocity of perturbations in turbulent channel flow. *Phys. Fluids A Fluid Dyn.*, 695(May 2013):695, 1993. doi:10.1063/1.858653.

Juan C. del Álamo and Javier Jiménez. Estimation of turbulent convection velocities and corrections to Taylor's approximation. *J. Fluid Mech.*, 640:5, 2009. doi:10.1017/S0022112009991029.

Ronald J. Adrian, C. D. Meinhart, and C. D. Tomkins. Vortex organization in the outer region of the turbulent boundary layer. *J. Fluid Mech.*, 422:1–54, 2000. doi:10.1017/S0022112000001580.

Theodore Theodorsen. Mechanism of turbulence. In *Proc. 2nd Midwest. Conf. Fluid Mech.*, 1952.

M. R. Head and P. Bandyopadhyay. New aspects of turbulent boundary-layer structure. *J. Fluid Mech.*, 107(-1):297, jun 1981. doi:10.1017/S0022112081001791.

Jigen Zhou, Ronald J. Adrian, and S. Balachandar. Autogeneration of near-wall vortical structures in channel flow. *Phys. Fluids*, 8(1):288, 1996. doi:10.1063/1.868838.

Jigen Zhou, Ronald J. Adrian, S. Balachandar, and T. M. Kendall. Mechanisms for generating coherent packets of hairpin vortices in channel flow. *J. Fluid Mech.*, 387:353–396, may 1999. doi:10.1017/S002211209900467X.

Adrián Lozano-Durán, Oscar Flores, and Javier Jiménez. The three-dimensional structure of momentum transfer in turbulent channels. *J. Fluid Mech.*, 694:100–130, feb 2012. doi:10.1017/jfm.2011.524.

a. a. Townsend. Equilibrium layers and wall turbulence. *J. Fluid Mech.*, 11(01):97, 1961.

A. E. Perry and M. S. Chong. On the mechanism of wall turbulence. *J. Fluid Mech.*, 119(-1):173, 1982. doi:10.1017/S0022112082001311.

Juan C. del Álamo, Javier Jiménez, Paulo Zandonade, and Robert D. Moser. Self-similar vortex clusters in the turbulent logarithmic region. *J. Fluid Mech.*, 561:329, aug 2006. doi:10.1017/S0022112006000814.

J. Nikuradse. *Untersuchung über die Geschwindigkeitsverteilung in turbulenten Strömungen.* PhD thesis, Göttingen, 1926.

W. D. Baines and E. Brundrett. The production and diffusion of vorticity in duct flow, 1964.

F. B. Gessner. The origin of secondary flow in turbulent flow along a corner. *J. Fluid Mech.*, 58 (01):1, mar 1973. doi:10.1017/S0022112073002090.

A. Melling and J. H. Whitelaw. Turbulent flow in a rectangular duct. *J. Fluid Mech.*, 78(02):289, apr 1976. doi:10.1017/S0022112076002450.

Asmund Huser and Sedat Biringen. Direct numerical simulation of turbulent flow in a square duct. *J. Fluid Mech.*, 257:65–95, 1992.

Hassan Raiesi, Ugo Piomelli, and Andrew Pollard. Evaluation of Turbulence Models Using Direct Numerical and Large-Eddy Simulation Data. *J. Fluids Eng.*, 133(2):021203, 2011. doi:10.1115/1.4003425.

Markus Uhlmann, Genta Kawahara, and Alfredo Pinelli. Traveling-waves consistent with turbulence-driven secondary flow in a square duct. *Phys. Fluids*, 22(8):084102, 2010. doi:10.1063/1.3466661.

Iehisa Nezu and Wolfgang Rodi. Experimental study on secondary currents in open channel flow. In *Proc. 21st Congr. IAHR, Melbourne, Aust.*, volume 2, pages 115–119, 1985.

Akihiro Tominaga, Iehisa Nezu, Kazuhiro Ezaki, and Hiroji Nakagawa. Three-dimensional turbulent structure in straight open channel flows. *J. Hydraul. Res.*, 27(1):149–173, jan 1989. doi:10.1080/00221688909499249.

P. Stearns. On the Current-Meter: Together with A Reason Why the Maximum Velocity of Water Flowing In Open Channels Is Below the Surface. *Trans. Am. Soc. Civ. Eng.*, XII(1):301–338, 1883.

H.J. Tracy and C.M. Lester. Resistance Coefficients and Velocity Distribution Smooth Rectangular Channel. Technical report, U.S. Geological Survey, 1961.

Kandula V. N. Sarma, P. Lakshminarayana, and N. S. Lakshmana Rao. Velocity Distribution in Smooth Rectangular Open Channels. *J. Hydraul. Eng.*, 109(2):270–289, 1983. doi:10.1061/(ASCE)0733-9429(1983)109:2(270).

Kandula V. N. Sarma, Boyina V. Prasad, and Avvari Krishna Sarma. Detailed Study of Binary Law for Open Channels. *J. Hydraul. Eng.*, 126(3):210–214, 2000.

Xingkui Wang, Zhao-Yin Wang, Mingzhong Yu, and Danxun Li. Velocity profile of sediment suspensions and comparison of log-law and wake-law. *J. Hydraul. Res.*, 39(2):211–217, 2001. doi:10.1080/00221680109499822.

Shu-qing Yang, Soon-keat Tan, and Siow-yong Lim. Velocity Distribution and Dip-Phenomenon in Smooth Uniform Open Channel Flows. *J. Hydraul. Eng.*, 130(12):1179–1186, dec 2004. doi:10.1061/(ASCE)0733-9429(2004)130:12(1179).

Iehisa Nezu and Hiroji Nakagawa. *Turbulence in Open-Channel Flows*. IAHR-Monograph, Balkema, Rotterdam, The Netherlands, 1993. ISBN 9054101180.

Iehisa Nezu. Open-Channel Flow Turbulence and Its Research Prospect in the 21st Century. *J. Hydraul. Eng.*, 131(4):229–246, 2005. doi:10.1061/(ASCE)0733-9429(2005)131:4(229).

L. M. Grega, T. Wei, R. I. Leighton, and J. C. Neves. Turbulent mixed-boundary flow in a corner formed by a solid wall and a free surface. *J. Fluid Mech.*, 294(-1):17–46, apr 1995. doi:http://dx.doi.org/10.1017/S0022112095002795.

T. Y. Hsu, L. M. Grega, R. I. Leighton, and T. Wei. Turbulent kinetic energy transport in a corner formed by a solid wall and a free surface. *J. Fluid Mech.*, 410:343–366, may 2000. doi:10.1017/S0022112099008125.

L. M. Grega, T. Y. Hsu, and T. Wei. Vorticity transport in a corner formed by a solid wall and a free surface. *J. Fluid Mech.*, 465:331–352, sep 2002. doi:10.1017/S0022112002001088.

T.G. G Thomas and J.J.R. J R Williams. Large eddy simulation of turbulent flow in an asymmetric compound open channel. *J. Hydraul. Res.*, 33(1):27–41, jan 1995. doi:10.1080/00221689509498682.

Madhu Sreedhar and Fred Stern. Large eddy simulation of temporally developing juncture flows. *Int. J. Numer. Methods Fluids*, 28(1):47–72, jul 1998. doi:10.1002/(SICI)1097-0363(19980715)28:1<47::AID-FLD698>3.0.CO;2-P.

Riccardo Broglia, Andrea Pascarelli, and Ugo Piomelli. Large-eddy simulations of ducts with a free surface. *J. Fluid Mech.*, 484:223–253, jun 2003. doi:10.1017/S0022112003004257.

Younghoon Joung and Sung-uk Choi. Direct numerical simulation of low Reynolds number flows in an open-channel with sidewalls. *Int. J. Numer. Methods Fluids*, (2):854–874, 2010. doi:10.1002/fld.

Atsushi Sekimoto, Genta Kawahara, K. Sekiyama, Markus Uhlmann, and Alfredo Pinelli. Turbulence- and buoyancy-driven secondary flow in a horizontal square duct heated from below. *Phys. Fluids*, 23(7):075103, 2011. doi:10.1063/1.3593462.

Genta Kawahara, Kyozo Ayukawa, Junji Ochi, Fumihiro Ono, and Eiji Kamada. Wall Shear Stress and Reynolds Stresses in a Low-Reynolds-Number Turbulent Square-Duct Flow. *Trans. Japan Soc. Mech. Eng. Ser. B*, 66(641):95–102, 2000. doi:10.1299/kikaib.66.95.

Atsushi Sekimoto. *Coherent Structures and Secondary Flow in Turbulent Square-Duct Flow*. Phd thesis, Osakai University, 2011.

Sandrine Hugues and Anthony Randriamampianina. An improved projection scheme applied to pseudospectral methods for the incompressible Navier–Stokes equations. *Int. J. Numer. Methods Fluids*, 521(February 1997):501–521, 1998.

Yoshiyuki Sakai and Markus Uhlmann. High-Resolution Numerical Analysis of Turbulent Flow in Straight Ducts with Rectangular Cross-Section. In Wolfgang E. Nagel, Dietmar B. Kroener, and Michael M. Resch, editors, *High Perform. Comput. Sci. Eng. '15*, pages 301–313. Springer International Publishing, Cham, 2016. doi:10.1007/978-3-319-24633-8_20.

David L. Brown, Ricardo Cortez, and Michael L. Minion. Accurate Projection Methods for the Incompressible Navier–Stokes Equations. *J. Comput. Phys.*, 168(2):464–499, apr 2001. doi:10.1006/jcph.2001.6715.

M Mohan Rai and Parviz Moin. Direct simulations of turbulent flow using finite-difference schemes. *J. Comput. Phys.*, 96:15–53, 1991.

P Haldenwang and G Labrosse. Chebyshev 3-D spectral and 2-D pseudospectral solvers for the Helmholtz equation. *J. Comput. Phys.*, (538):115–128, 1984.

C. Canuto, M. Y. Hussaini, A. Quarteroni, and T. A. Zang. *Spectral Methods*. Scientific Computation. Springer Berlin Heidelberg, Berlin, Heidelberg, 2006. ISBN 978-3-540-30725-9. doi:10.1007/978-3-540-30726-6.

Hao Zhang, F. Xavier Trias, Andrey Gorobets, Yuanqiang Tan, and Assensi Oliva. Direct numerical simulation of a fully developed turbulent square duct flow up to Re_\tau=1200. *Int. J. Heat Fluid Flow*, 54:258–267, 2015. doi:10.1016/j.ijheatfluidflow.2015.06.003.

O. C. Jones. An improvement in the calculation of turbulent friction in rectangular ducts. *J. Fluids Eng.*, 1976.

Shigeo Kida and Hideaki Miura. Swirl condition in low-pressure vortices. *J. Phys. Soc. Japan*, 67 (7):2166–2169, 1998.

Paolo Orlandi. Vortex dipole rebound from a wall. *Phys. Fluids A Fluid Dyn.*, 2(8):1429, 1990. doi:10.1063/1.857591.

Ryuichi Nagaosa. Direct numerical simulation of vortex structures and turbulent scalar transfer across a free surface in a fully developed turbulence. *Phys. Fluids*, 11(6):1581, 1999. doi:10.1063/1.870020.

Tatsuya Nakatsuji. *Large-scale structures and secondary flow in turbulent square-duct flow*. Msc. thesis, Osaka University, 2012.

Thibaut Nowakowski. *Localized turbulence in duct flow : A numerical investigation*. Msc. thesis, Ecole Polytechnique, 2007.

Jinhee Jeong and Fazle Hussain. On the identification of a vortex. *J. Fluid Mech.*, 285(-1):69–94, apr 1995. doi:10.1017/S0022112095000462.

J. A. Lehew, M. Guala, and B. J. McKeon. Time-resolved measurements of coherent structures in the turbulent boundary layer. *Exp. Fluids*, 54(4), 2013. doi:10.1007/s00348-013-1508-4.

Adrián Lozano-Durán and Javier Jiménez. Time-resolved evolution of coherent structures in turbulent channels: characterization of eddies and cascades. *J. Fluid Mech.*, 759:432–471, oct 2014. doi:10.1017/jfm.2014.575.

Ricardo Vinuesa, Azad Noorani, Adrián Lozano-Durán, George K. El Khoury, Philipp Schlatter, Paul F Fischer, and Hassan M. Nagib. Aspect ratio effects in turbulent duct flows studied through direct numerical simulation. *J. Turbul.*, 15(10):677–706, oct 2014. doi:10.1080/14685248.2014.925623.

Takashi Ishihara, Yukio Kaneda, M. Yokokawa, K. Itakura, and a. Uno. Small-scale statistics in high-resolution direct numerical simulation of turbulence: Reynolds number dependence of one-point velocity gradient statistics. *J. Fluid Mech.*, 592:335–366, 2007. doi:10.1017/S0022112007008531.

Sergio Hoyas and Javier Jiménez. Reynolds number effects on the Reynolds-stress budgets in turbulent channels. *Phys. Fluids*, 20(10):101511, 2008. doi:10.1063/1.3005862.

U. Ehrenstein and R Peyret. A Chebyshev collocation method for the Navier-Stokes equations with application to double-diffusive convection. *Int. J. Numer. Methods Fluids*, 9(4):427–452, apr 1989. doi:10.1002/fld.1650090405.

L Quartapelle and F Valz-Gris. Projection conditions on the vorticity in viscous incompressible flows. *Int. J. Numer. Methods Fluids*, 1(January):129–144, 1981.

Weinan E and Jian-Guo Liu. Vorticity Boundary Condition and Related Issues for Finite Difference Schemes. *J. Comput. Phys.*, 124(2):368–382, mar 1996. doi:10.1006/jcph.1996.0066.

Olivier Botella. On the solution of the Navier-Stokes equations using Chebyshev projection schemes with third-order accuracy in time. *Comput. Fluids*, 26(2):107–116, feb 1997. doi:10.1016/S0045-7930(96)00032-1.

Jie Shen. Hopf bifurcation of the unsteady regularized driven cavity flow. *J. Comput. Phys.*, 245 (199 1):228–245, 1991.

Olivier Botella and R Peyret. Computing singular solutions of the Navier-Stokes equations with the Chebyshev-collocation method. *Int. J. Numer. Methods Fluids*, 36(2):125–163, may 2001. doi:10.1002/fld.121.

Dissertationsreihe am Institut für Hydromechanik
Karlsruher Institut für Technologie (KIT)

ISSN 1439-4111

Die Bände sind unter www.ksp.kit.edu als PDF frei verfügbar oder als Druckausgabe bestellbar.

2004/1 **Volker Weitbrecht**
Influence of dead-water zones on the dispersive mass-transport in rivers. 2004
ISBN 3-937300-07-4

2004/2 **Christof Hinterberger**
Dreidimensionale und tiefengemittelte Large-Eddy-Simulation von Flachwasserströmungen. 2004
ISBN 3-937300-18-X

2005/1 **Cornelia Frank**
Wirksamkeit von dünnen Windschutzstreifen auf Sockelwällen in luv- und leeseitiger Anordnung. 2005
ISBN 3-937300-55-4

2005/2 **Carl Friedrich von Carmer**
Shallow turbulent wake flows: momentum and mass transfer due to large-scale coherent vortical structures. 2005
ISBN 3-937300-63-5

2005/3 **Muhammad Ikhwan**
Investigation of flow and pressure characteristics around pyramidal buildings. 2005
ISBN 3-937300-72-4

2005/4 **Herlina**
Gas transfer at the air-water interface in a turbulent flow environment. 2005
ISBN 3-937300-74-0

2006/1 **Manual García-Villalba Navaridas**
Large eddy simulation of turbulent swirling jets. 2006
ISBN 3-86644-015-4

2006/2 **Klaas E. Heinrich**
Hydraulische Steuerung einer Wasser-Alkohol-Zirkulation mittels Grundwasser-Zirkulations-Brunnen. 2006
ISBN 3-86644-055-3

2006/3 **Tobias Bleninger**
Coupled 3D hydrodynamic models for submarine outfalls. Denvironmental hydraulic design and control of multiport diffusers. 2006
ISBN 978-3-86644-160-6

2006/3 **Patrick Heneka**
Schäden durch Winterstürme – das Schadensrisiko von Wohngebäuden in Baden-Württemberg. 2006
ISBN 978-3-86644-117-0

2007/1 **Maria Eletta Negretti**
Hydrodynamic instabilities and entrainment in two-layer stratified flows down a slope. 2007
ISBN 978-3-86644-170-5

2007/2 **Gregor Kühn**
Untersuchungen zur Feinsedimentdynamik unter Turbulenzeinfluss. 2007
ISBN 978-3-86644-205-4

2008/1 **Martin Detert**
Hydrodynamic processes at the water-sediment interface of streambeds. 2008
ISBN 978-3-86644-244-3

2008/2 **Christof Gromke**
Einfluss von Bäumen auf die Durchlüftung von innerstädtischen Straßenschluchten. 2008
ISBN 978-3-86644-339-6

2009/1 **Ekkehart Bethge**
Risikoberechnung zum Schadstoffeintrag aus Hochwasserretentionsräumen in einen Grundwasserleiter. 2009
ISBN 978-3-86644-426-3

2012/1 **Clemens Chan-Braun**
Turbulent open channel flow, sediment erosion and sediment transport. 2012
ISBN 978-3-86644-900-8

2015/1 **Todor Doychev**
The dynamics of finite-size settling particles. 2015
ISBN 978-3-7315-0307-1

2016/1 **Aman Ghebremichael Kidanemariam**
The formation of patterns in subaqueous sediment. 2016
ISBN 978-3-7315-0532-7

2019/1 **Yoshiyuki Sakai**
Coherent structures and secondary motions in open duct flow. 2019
ISBN 978-3-7315-0790-1